D1286817

Also by Marc Lappé

Of All Things Most Yielding
(with John Chang McCurdy)

GENETIC POLITICS

The Limits of Biological Control

MARC LAPPÉ

Simon and Schuster
NEW YORK

Published by Simon and Schuster
A Division of Gulf & Western Corporation
Simon & Schuster Building
Rockefeller Center
1230 Avenue of the Americas
New York, New York 10020

Designed by Stanley S. Drate
Manufactured in the United States of America
1 2 3 4 5 6 7 8 9 10

Library of Congress Cataloging in Publication Data

Lappé, Marc.
Genetic politics.
Includes bibliographical references and index.
1. Human genetics—Social aspects. 2. Genetics—Social aspects. 3. Genetic engineering—Social aspects. I. Title
QH438.7.L36 575.1 79-16403

ISBN 0-671-22546-4

For Nichol—
who believed in me

Acknowledgments

My heartfelt thanks to those who gave their support and encouragement during the preparation of this book—my family, who gave me unstinting faith through the period when work was most difficult; my friends who urged me to take the project to completion; and those who offered freely of their comments, advice and criticism.

I wish to thank especially Cary Fowler, Peggy Sloane and Deanna Beeler, who helped me through a long New York City winter; and Sarah Fowler and Alma Chervin, my research assistants, whose enthusiasm and personal commitment to the ideas expressed in the book fueled my efforts and research.

Special thanks also to Phil Reilly, Tom Fanning and John Siciliano, whose papers, theses and comments provided a fertile ground for many of my ideas and concepts. I also wish to acknowledge the stimulation of my colleagues at the Hastings Center, Peter Steinfels, Will Gaylin, Dan Callahan, Tabitha Powledge, and particularly Art Caplan, who challenged the major thesis of the book and so shaped its form. The shared experience of the Center Fellows who served on the Genetics Research Group that I directed proved an invaluable resource for the material in the book that is at the interface of genetics and ethics. I am also extremely grateful for the selfless review and critique of drafts of the book by Ted Friedmann and Phil Reilly.

I want to express my enduring appreciation to Lloyd Linford, who encouraged me to revive the book, and whose brilliant editing gave the book whatever stylistic grace it now has. My gratitude also to Alice Mayhew for her encouragement and support, and for keeping the book alive in her mind over four years of writing and development.

My thanks to those teachers and mentors who showed me the fallibility of simplistic thinking and the complexity of causation, S. Meryl Rose and Richmond T. Prehn. And finally, my warmest appreciation goes to my wife, Nichol Lovera, whose love and faith made it possible to renew a dream after I had set it aside.

CONTENTS

1

Introduction

Contrary to the lessons taught in most genetic textbooks, the origins of genetic thinking extend well before the life and times of an otherwise obscure monk named Gregor Mendel (1822–84).

Even before genetics was grounded in biological fact, people used rule-of-thumb breeding methods in the practice of rudimentary agriculture and animal husbandry. Constant exposure to the patterns of inheritance in nature must certainly have impressed the primitive mind. The recognition that "like produces like" is an elementary sign of a dawning genetic consciousness, one that is expressed also in preoccupation with genealogy, both human and divine. Such a belief has spawned some of the most oppressive social policies known to human societies. Genetic awareness must also have been critical in the establishment of the kinship systems that ensured exogamy (marrying outside the clan) or, very much more rarely, endogamy (marrying within it).

Claude Levi-Strauss has described the penchant of the "savage" mind for ordering the natural world into familylike relationships. Evolutionary relationships were intuited and characterized by "primitive" peoples like those of Papua, New Guinea with uncanny accuracy. In these groups, closely related birds of paradise were called "brother" and "sister," more distant ones "cousin," and so forth, closely reflecting the actual taxonomic or evolutionary distance between species.

Forerunners of our modern understanding of genetics can be traced to the early eighteenth century in the works of Diderot and Maupertuis, who catalogued and charted patterns of familial resemblance with scientific precision. By the early 1800s, English physicians like Joseph Adams had identified the peculiar pattern of inheritance that marked the course of hemophilia through generations of affected male children and seemingly left females completely unaffected. We now recognize in this pattern a distribution of genes from the female line to the male through a mechanism of "X-linked" inheritance. In this form of inheritance, a gene (or genes) found on the X chromosome is responsible for an effect in males, who receive it from their mothers.

We owe our understanding of the rudiments of the three other major forms of inheritance—dominant, recessive and multifactorial—to Gregor Mendel. In the period between 1858 and 1865, when he finally published his work in the *Journal of the Medical Society of Brno* (in what is now Czechoslovakia), Mendel prepared the soil for our thinking about simple genetic systems.

Working assiduously on his pea plants to avoid characteristics that were equivocal, and hence not subject to easy enumeration, Mendel selected some twenty-one features that could be followed from one generation to the next with almost total accuracy.

Mendel concentrated only on clearly visible and enumerable features, like the texture and color of the peas or flowers, or the height of the pea plants. By careful testing,

Mendel was able to eliminate features that appeared to blend or to be intermediate in height or color or shape when he crossbred two plant lines. Instead, he chose purebred plants whose features were constant and unvarying, so that when he crossed them the hybrid offspring would resemble one parent almost entirely.

Mendel published his results in a paper entitled *Experiments in Plant Hybridization,* reflecting more than eight years of systematic study. He was able to show that essentially all the characteristics that he had *chosen* to study (and this is important) were distributed in the offspring of his hybrid plants with mathematical precision. Mendel's genius appeared to be not only in recording the characteristics that had been found in both plants with mathematical accuracy, but also in counting enough plants to derive *ratios* in the proportion of variants resulting from hybrid crosses.

Whenever Mendel crossed purebred plants with the clear-cut features that he had chosen, one characteristic would recede from view while the other dominated to give a common parentlike appearance of all hybrids. In his own words, "Those characters that are transmitted entire or almost unchanged in hybridization and therefore in themselves constitute the characters of the hybrid are termed dominant, and those that become latent in the process [are termed] recessive." He called the features that were hidden *recessive.* The fact that the recessive character reappeared with uncanny precision in the crosses of his hybrids led Mendel to conclude that he was dealing with *physical units,* each one capable of producing and reproducing a specific character. In this way, he reasoned, one physical "bit" (only in 1913 did William Bateson call these bits *genes)* was responsible for one character.

In a typical experiment, Mendel crossed tall and short plants to get tall-appearing hybrids. Upon crossing the hybrids, he produced 787 tall and 277 short out of 1,064 plants, for a ratio of 2.84 to 1. He made the conceptual leap to round this figure to 3 to 1, while at the same time reasoning that the only simple way to explain his results was to assume that each hybrid pea plant produced two factors, one for tall and the other for short. Moreover, Mendel reasoned, these factors were apparently distributed in the germ cells in equal numbers, so that on the average, half of a large number of offspring of two hybrid plants would get one tall factor (T) and one short factor (t); one quarter would get two tall factors; and the remaining quarter two short factors (see diagram). Since, in this example, tall overrode, or *dominated*, short, three fourths of his plants would be tall and one fourth short, for a ratio of tall to short plants of 3 to 1.

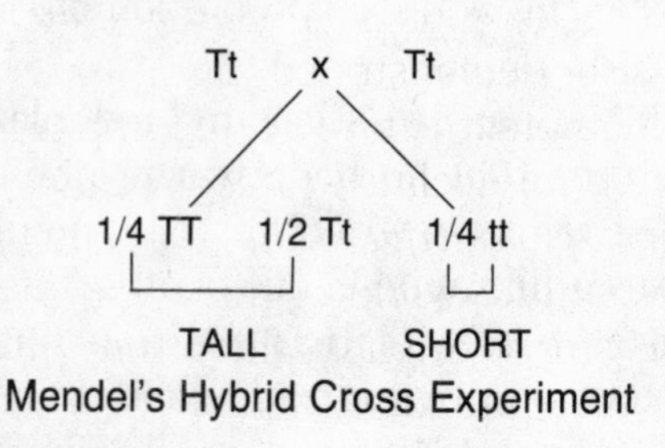

Mendel's Hybrid Cross Experiment

Mendel's results have seemed totally convincing and appealing in their simplicity, yet his mentor, Karl Nägeli, believed that his data were trivial and not capable of being generalized—largely, many now feel, because Nägeli's favorite botanical subjects did not show the consistent results that Mendel's carefully selected pea varieties did. As to the response of the general scientific community, many believe that Mendel contributed to his own neglect by not publishing in a more widely read journal. But an often overlooked peculiarity in Mendel's procedures may have been responsible.

Mendel's greatest contribution was in his use of statistics to generalize from individual data; to do this, of course, he needed large numbers. But in an era when there was little or no appreciation for imprecision or chance variation, Mendel may not have had a "feel" for variability. We now believe that Mendel was so concerned about proving his case that he carefully selected and unconsciously biased his data to reflect his expectations. As a result of a reexamination of Mendel's work performed in 1934 by the famous statistician Ronald A. Fisher, it is clear that Mendel's ratios were too exact to be explained by chance. The statistical probability of getting the reported closeness of fit between raw data and the numerical ratios that Mendel reported was about 1 chance in 14,000—a finding that strongly suggests that Mendel knew what he was looking for.

Mendel also demonstrated how numerous inherited factors could interact to produce the appearance of "blending," although he declined to explore the nature of what we now know as *multifactorial* inheritance in depth.

But the scientific world of the mid-nineteenth century was not ready for so dramatically simple an explanation for the truism of like producing like, or for Mendel's elegantly straightforward explanation for the bald fact of the reemergence of previously hidden characteristics after hybrid plants—or animals—were crossed. Yet, even while the scientific community ignored his findings, the public proved receptive and eager for new interpretations of the role of inheritance in human affairs.

One Englishman in particular, Francis Galton (1822–1911),was a central figure in the popularization of familial patterns of inheritance. It was Galton who first deduced that parents contribute 50 percent to the over-all genetic makeup of an individual; the four grandparents 25 percent; the third generation of ancestors 12.5 percent, and so on. In 1869 Galton published his *Hereditary Genius*, a massive work that purported to show, through analyses of numerous men of famous families, that personal qualities were no less hereditary than were physiognomy and physical stature. In popular media, Galton claimed that skill, talent and good character were subject to the laws of heredity. In 1883 he coined the word "eugenics" to describe the systematic study and application of the ideas of heredity to humans, with the objective of improving (or at least stopping the degradation of) human genetic stock. His ideal of race improvement was predicated on developing a sound theoretical foundation for eugenics. But the scientific and statistical tools available at the time were unequal to the task.

Galton buttressed his case with the family studies of the peers in the House of Lords, the relatives of eminent jurists and scholars who lived between 1660 and 1860, the lineage of Queen Victoria, and other eminent examples of hereditary excellence, which he explored in his *Hereditary Genius*. He even made the first inroads into the mathematical complexities of the inheritance of stature and other multifactorial traits. Perhaps because he concentrated on such complex traits, he did not recognize the existence of single-gene–determined characters.

In the spring of 1900 occurred an event that demonstrated the truism of historical readiness for an idea "whose time has come." Fully *six* papers were published, by three independent investigators, all reporting the rediscovery of Mendel's lost work on hybrid inheritance. Hugo de Vries of Holland, Karl Correns of Austria and Erich von Tschermak of Germany all repeated Mendel's studies and acknowledged, somewhat belatedly, their indebtedness for the success of their studies to the precedent set by Mendel, fully thirty-four years earlier.

For the first two or three decades after the rediscovery of "Mendel's Laws" (as Karl Correns called them), many scientists were optimistic that much of human misery and disease, as well as the more positive characteristics of "natural ability" and "intelligence" that Francis Galton had studied at the end of the nineteenth century, would lend themselves to simple inheritance patterns. Only a few skeptics like J. B. S. Haldane and R. A. Fisher (who were to unite Mendelian genetics and population genetics only a few years later) argued that many human factors appeared to be beyond reduction to single genes. But the synthetic view that many complex conditions could be explained only on the basis of many different genes acting in concert with the environment, did not hold much sway for geneticists and social planners in the early 1920s and 1930s.

The prevailing view of the early 1900s was that one gene pair was responsible for one character in humans just as in peas. Skull shape, stature, eye and hair color, and the shape of the nose were all declared to be inherited according to Mendel's Laws. And some scientists believed that, like hybrid peas, racial mixes that appeared to produce harmonious hybrids *(sic)* like mulattoes, could in later generations break down to allow disharmonious characteristics to reappear. During this period, tuberculosis, imbecility, syphilis, insanity, prostitution, alcoholism, and even pellagra (a nutritional-deficiency disease)* were also attributed to single genetic factors.

*Pellagra, a severe and debilitating constitutional disorder accompanied by mania and severe reddening of the skin, was believed to have a genetic basis because it occurred preferentially among the rural poor in the South, particularly in families which subsisted on corn as the major dietary staple. It was a major health problem of the early 1900s, affecting tens of thousands of "poor white trash" and blacks. For over a decade after Joseph Goldberger of the US Public Health Service (circa 1914) firmly established pellagra's dietary origins in a deficiency of tryptophan and/or the B vitamin nicotinic acid, the genetic hypothesis was kept alive. Largely through the efforts of Charles Benedict Davenport, the head of the Eugenics Record Office of Cold Spring Harbor, New York, Goldberger's studies were suppressed and environmental remedies postponed until after the Great Depression. In the interim, over 80,000 people died of a simple, treatable nutritional deficiency because a genetic idea held sway over more rational interpretations.

The American Life Insurance Medical Directors recorded these factors as well as several bona fide single-gene diseases in their *Proceedings* between 1900 and 1950 as a guide for underwriting policies. And governing bodies, most notably the United States Congress, used the presumption that genetic factors lay behind certain undesirable traits as one of the grounds to exclude aliens of a particularly "undesirable" origin.

Ten years after the rediscovery of Mendel's work, the American psychologist Henry Goddard imported the Binet intelligence test from France and used it to grade feeble-minded children. Although Binet designed the test to help rescue otherwise intelligent children who were warehoused with the mentally impaired because of their physical handicaps, others found in the IQ tests a means to show that "low intelligence" was characteristic of virtually every socially undesirable group. Public policy, not Mendel, made it essential to link genetics with human disease and disability. This link proved one way to ensure a "rational" reason for painting all ethnically related persons with the brush of genetic taint. A "Committee to Study and Report on the Best Practical Means to Cut Off the Supply of Defective Germ Plasm in the American Population" filed its reports in 1912 and concluded that eugenic sterilization was socially justified.

In 1913 geneticist H. E. Walker called for the awakening, among Americans, of a "eugenic conscience" that would lead to the implementation of social policies to improve mankind. As a partial list of what would be needed, Walker itemized control of immigration, more discriminating marriage laws, and neuterization or sterilization "where necessary." Restricting immigration, in Walker's view, was the only sure way to prevent the ingress of "potentially bad germoplasm," which "might wreak havoc on future generations." To illustrate his point, he listed among those "caught in the sieve" at Ellis Island in 1908: 65 idiots, 121 feeble-minded, 184 insane, 3,741 paupers, 2,900 persons with contagious diseases, 53 with tuberculosis, 136 criminals and 124 prostitutes. By the early 1920s, Congress had adopted the Johnson Act, a statute that limited immigration to the United States based in part on this kind of genetic rationale.

Between 1913 and 1930, a kind of pseudo-Mendelian genetics was used by eugenicists like Paul Popenoe, editor of the *Journal of Heredity*, and R. H. Johnson to prove that "weak heredity" accounted for the high rates of infant mortality among the poor, and that their inadequate income and low standard of living was largely the consequence of poor heredity.

Before World War II, programs based on unscientific views of genetics were still readily translated into public policy. Sterilization in particular came to be regarded as a legitimate tool to control deviance and reduce the future burden of the handicapped on a progressive society.

In 1907, Indiana became the first state to enact a sterilization law. Over the next twenty-five years, forty-four more states passed similar legislation, and certain crimes and many forms of mental illness were found to be justifiable causes for compulsory sterilization. By 1935 twenty of the states had enacted eugenic-sterilization laws, 20,000 Americans had undergone eugenic sterilization, and the Supreme Court had upheld their constitutionality in the celebrated case of *Buck* vs. *Bell,* in which a backwoods mother, Carrie Buck, was ordered sterilized under a Virginia statute. For many social planners, eugenic sterilization struck a blow at the heartwood of defect and deviance that were the impediments to social programs.

In formulating their programs, advocates of sterilization drew from the nascent science of genetics only those discoveries and ideas that bolstered a simple Mendelian picture of the primacy of hereditary factors. Those who were most active in pressing for eugenic reform selectively used new developments in the theory of heredity and, by necessity, simplified and distorted the then-burgeoning knowledge about the complex basis for most retardation, alcoholism and epilepsy.

So much as been made of such "abuses" of genetics that it is important to understand that these examples represent not the excesses of a few zealots, but often reflected the concern of a whole society (at least in the early 1920s) that American social progress be the hallmark for the world. Today the specter of genetic abuse is much reduced and softened by a sophisticated understanding of genetics. But that knowledge is often not accessible to the politically powerful.

Today, even though sterilization of the institutionally retarded is under extremely tight state and federal control as the result of restrictions on the use of public funds to sterilize minors or nonconsenting adults, there are still vestiges of ignorance about the purpose that such sterilization could serve. Thus, many persons, including physicians with whom I have communicated, are still concerned that our failure to sterilize the mentally defective will result in an increase in the incidence of mental retardation. Such a view is reflected in the tone and content of a 1974 article in *Fortune* magazine entitled "What Science Can Do About Hereditary Disease," in which the author encourages his readers to reconsider eugenics.

What do we know about the genetics of inheritance today? There are presently an astonishing number of human disorders associated with "Mendelian" (that is, single-gene) inheritance. An annual catalogue edited by Victor McKusick lists almost three thousand disorders known or suspected to be linked to single-gene defects.

Considering their number, the aggregate health impact of this protean list of disorders is surprisingly small. With the exception of such diseases as cystic fibrosis, sickle-cell anemia and Tay-Sachs disease, the incidence of any one single-gene disorder among the American population rarely exceeds 1 in every 15,000 persons. And taken together, the three-thousand-or-so single-gene disorders affect probably no more than thirty-six newborns out of every ten thousand births, according to a review published in the November-December 1977 issue of the *American Scientist* by San Francisco geneticists Charles Epstein and Mitchell Golbus. These same authors indicate that only about one and a half times that number, or slightly more than 0.5 percent of newborns, will be born with clinically important chromosomal abnormalities.

Overall, researchers estimate that some 2.5 to 3 percent of all newborns are born with a birth defect. At present, only about half of these are clinically apparent at birth, and an even smaller number associated with a single gene defect, with the upper limit set at 1.2 percent.

While these figures will undoubtedly rise as a greater appreciation of the genetic basis for disease and minor variations in the biological makeup of humans become better understood, the *major* patterns of the genetic contributions to human disease are fairly well mapped out. (By comparison with virtually every other well-studied life form, knowledge of the precise location of genes on the human chromosomes is pitifully small.) The picture that emerges is one of a profusion of relatively rare gene-based disorders, with the aggregate effect of other potentially damaging genes remarkably well dampened and buffered by the rest of the genetic makeup of most persons. Indeed, with the exception of dominantly inherited conditions, few known deleterious genes make their presence felt in the carrier individual. (More about this later.)

What we do know is that most genetic disorders can be listed under four broad categories: (1) autosomal recessive; (2) dominant; (3) X-linked; and (4) multifactorial. The first three account for roughly 25, 6, and 5 disorders, respectively, in every 10,000 newborns; while the fourth category is responsible alone for disorders in up to 165 newborns in every 10,000. Since we shall be referring to disease entities or characteristics found in each of these categories, it is useful to set them in perspective and to understand their patterns of inheritance.

First, it is well to keep in mind that the genes we are describing are arranged along the length of some 46 chromosomes. Forty-four of these chromosomes are called "autosomes" and are arranged into 22 pairs. Each parent provides one set. The remaining two chromosomes are the sex chromosomes. They consist of the X and the Y; normally, males have one of each, females have two X's. Each parent provides a single sex chromosome, which combined with the 22 autosomes ensures that each contributes half of the full complement of 46 chromosomes.

Autosomal recessive conditions number 1,117 in McKusick's 1978 catalogue. In autosomal recessive inheritance, the individual receives *two genes* that are found at the same position on a common parental autosome. The combined effect of having received the same gene from each parent is deleterious only when that gene produces a defective product. A defective product has little or no biological activity or usefulness. Thus, the adverse effects of commonly studied autosomal recessive conditions like phenylketonuria (PKU) or galactosemia generally result from the presence of one kind of defective product; in this case an enzyme. In the case of PKU it is an enzyme needed to utilize an amino acid called phenylalanine that is faulty. In the case of galactosemia it is another enzyme needed to break down a sugar called galactose. (These two enzymes are called "phenylalanine 4-hydroxylase" and "galactokinase" respectively.) For these and the vast majority of other autosomal recessive diseases, many of which are known as "inborn errors of metabolism," most of the physiological damage results from the progressive accumulation of intermediate substances that get dammed up behind the blocked enzymatic reaction.

As in pea plants, recessive genes can be carried in a parent without making their presence known through any clinically visible signs. But if two parents carry the same recessive gene, as in Mendel's pea hybrids, one fourth of their children (on the average) will get a double dose of the responsible genes and will be affected by the deleterious effect of having faulty or nonfunctional biochemicals.

Sickle-cell anemia, cystic fibrosis and Tay-Sachs disease are all produced by the interbreeding of carriers of the same recessive gene. For these, as with all other recessive diseases, the actual number of people who carry a single dose of the responsible gene is surprisingly large. For instance, about one in twelve blacks actually carries the gene for sickle-cell anemia, while on the average only one in every 576 newborns will get a double dose. (There is about one chance in 144 of two such people marrying— $1/12 \times 1/12$—and another one chance in four of their having an affected child—$1/12 \times 1/12 \times 1/4 = 1/576$.) Similarly, about one in sixty whites carries the major gene responsible for PKU, while only about one in every 14,000–15,500 is affected; about one in thirty Jews of Eastern European ancestry carries the gene for Tay-Sachs, and only one in about every 3,600 newborns expresses the disease, and so on.

All told, population geneticists estimate from figures like these that we must each carry somewhere between one and ten potentially deleterious autosomal recessive genes—with some of us having many truly lethal genes in the recessive form, and a few of us having none.

By comparison, *autosomal dominant* conditions require only that a single gene be inherited from either parent for the symptoms of a disorder or disease process to become manifest. In part because they are so much more likely to be detected in the population at large, 1,489 dominantly inherited disorders are presently catalogued. Achondroplasia (dwarfism) and Huntington disease (a late-appearing, neurologically destructive disease of which more will be said later) are but two of the many individually rare, but nonetheless significant dominant disorders. Fully one half of the affected person's offspring may be expected to inherit the same gene that caused the parent's disability.

Characteristically, the most common dominant diseases either occur late in life or produce only mildly disfiguring or disabling effects, since by and large it is only through the survival and reproduction of an affected person that the gene is perpetuated. (Any dominant condition can also occur at random as the result of a spontaneous change in the normal gene, known as a *mutation,* which in humans characteristically occurs only about once in every million genes each generation, or slightly less than 1 new mutation per gamete—egg or sperm).

X-linked conditions, of which some two hundred and five are listed in the McKusick catalogue, are like dominant conditions in that they trace their expression to the effects of a single gene, but in this instance the gene is on the female or X chromosome. While there are both dominant and recessive X-linked disorders, only the latter find their expression characteristically in males. The reason for this otherwise inexplicable affinity becomes clear if you consider that only males get a single X chromosome. For a female, the presence of another X chromosome with the normal gene works to negate any adverse consequences of a defective gene, just as does a normal autosomal gene for a carrier of a single deleterious recessive gene.

Thus, if a male happens to get the mother's X chromosome that has the defective recessive gene on it (something that happens half the time for male offspring of the carrier woman), he is unprotected, since the Y chromosome does not bear any genes in common with the X. A daughter receiving a similarly defective X-linked gene will usually escape unscathed in terms of the deleterious health effects of a defective gene, since she will almost always receive a full complement of normal genes on the X chromosome that she receives from her father. Hemophilia, of which two types, A and B, are most commonly known, is a classic example of X-linked, recessive inheritance.

A critical distinction between such X-linked recessive disorders and common autosomal recessive disorders is that X-linked ones show considerably more variability in their expression. In this sense, conditions like classic hemophilia resemble many dominantly inherited conditions in that they affect some persons very severely and others hardly at all. In the case of the bleeder's disease, as hemophilia is commonly known, only slightly more than half of the boys with the A form of the disease have a severe problem with bleeding episodes; as do fewer than 40 percent of those with the B form of the disease. These boys, on the average, have less than 2 percent of the amount of a critical clotting factor that normal boys have in their blood serum. The remainder have higher and, in some cases, near-normal levels.

Multifactorial disorders are by far the most prevalent, affecting about 1.2 to 1.4 percent of all births. As the name implies, multifactorial disorders are characterized by having many different factors in their origins and genesis, some genetic and some environmental.

The precise contribution of the different genes to any one multifactorial disease is as yet poorly understood. Conditions like diabetes, neural-tube defects, congenital heart malformations and hypertension all appear to be multifactorial in origin, as determined by careful family and twin studies. Because such multifactorial disorders, particularly clubfoot, neural-tube defects and heart defects are so common at birth, they have come under increasingly urgent scrutiny. Not least among the confounding findings is the fact that only rarely is a neural-tube defect like anencephaly (the failure of completion of the brain) or spina bifida (an opening at a place along the spinal column) found in both individuals of a pair of genetically identical twins. Results like these suggest the effects of powerful nongenetic influences, probably of environmental origin.

To appreciate how much new information of this kind we have had to incorporate over just the last twenty years, it is worth while to review the rapidity with which genetic knowledge has expanded. According to Victor McKusick's catalogue, the number of recorded disorders known or suspected to have a genetic origin jumped from 412 in 1958 to 1,545 just ten years later. By 1978 the number had risen to 2,811. With another 800–1,000 new disease entities associated with single genes expected before the end of the decade, the doubling time for discovery of new genetic diseases appears to be ten years or less. Thus, by the late 1980s we may expect to be faced with decisions that will have to take cognizance of more than *six thousand* different single-gene disorders!

Since 1908–9 when Sir Archibald Garrod first conceived the idea of genetically determined human metabolic diseases, the number of "errors of metabolism" detectable in the newborn has shown a similarly rapid increase. In the 1940s only a handful of such special newborn genetic diseases could be detected. By 1960 this number had reached twenty or so. And by the early 1970s, we could list over one hundred, with dozens of new entities being discovered yearly.

Our ability to detect genetically based disorders of metabolism or development *before* birth was the most recent frontier to be crossed. Prenatally detected disorders too have now joined this exponential rate of increase. Today, over sixty different genetic diseases can be ascertained through prenatal diagnosis, in time to provide critical information about the likely quality of life of a prospective human being, for parents, and less directly, for society, to make decisions as to whether to continue or to terminate a pregnancy.

And an even broader array of genetic and statistical tests are at hand today to measure the possible degree to which genes underlie the most significant causes of human disease and disability. We are on the threshold of uncovering the genes that predispose some people to mental health and others to mental illness. We know of genetic *markers* that signal the coming of serious arthritic conditions in some people and coronary artery disease in others. And some scientists believe we are close to knowing the genetic basis for the eventuality of succumbing to occupational disease or disability. These last possibilities suggest that genetics can radically change our lives. And procreative decisions that we previously made in blind ignorance have now been stripped of their veil of uncertainty. Through a genetic window, we can know with uncanny, sometimes disturbing accuracy the most intimate facts of the legacy brought by our spouse's background, and the quality of life we may expect for our children. And some of these facts can be known before birth or conception.

Over 285 different clinics and genetic units at hospitals and medical centers offered genetic counseling in 1977 to thousands of persons, and many more centers are expected. Over twenty million people have been screened for genetic diseases like PKU and sickle-cell anemia.

What all this means is that through genetic knowledge we seem to be on the verge of a new dimension of self-knowledge. Because of what we can know about genetics, we suddenly live in a world with infinitely more possibilities about where we came from and where we may be going. We have a brand-new knowledge base about the biological basis for individual differences in susceptibility to disease and perhaps ultimately about why we are all so different in body and mind.

As a result of advances in genetic knowledge, we live in a universe with infinitely more choices, a universe where many more judgments will have to be made—judgments as to where genetics rightfully affects our choices of spouse or profession, or where we should compensate (or penalize) one ethnic group for an inherited disability and not another. Because of the importance of these decisions, we must look critically on what it is that genetics can actually explain and where, perhaps, we have strained our projections of what genetics can do.

The roots of our expectations for the immense predictive power of genetics are readily apparent to any biological researcher who worked in the 1960s. I can remember the boundless optimism that fueled my research in transplantation biology in the late 1960s. There probably wasn't an immunology researcher who didn't believe that if we could just expand our knowledge of genetics far enough, we would be able to develop a complete understanding of why and how one animal recognized the tissues of another as "foreign." It wasn't so much that genes couldn't be counted on to play the core role in setting the limits to an animal's reactions, but we expected genetics to explain *everything!* In time we had to temper our expectations.

With the exception of identical twins, the most assiduous matching of tissue types from one person to another has never proved to be quite enough to assure the permanent success of a tissue graft, unless at the same time we knocked out the immunologic defenses of the intended recipient. Nongenetic factors, such as whether the graft was taken from a live donor or from a cadaver and whether the graft had been perfused to leach out all its white blood cells, appeared to be as important as genetic factors in predicting the success of grafts from unrelated donors. As immunological investigations continued, a host of these seemingly nongenetic factors were grudgingly recognized.

In my own work, I found that some animals might accept a genetically different graft indefinitely, while others with the same genetic background might not—even after they had received identical courses of immunization.* And small grafts on one side of the same animal might be destroyed while larger ones on the other side were accepted for days and even weeks longer. An immunized mother (in mice, of course) could destroy a graft handily while quietly nurturing a fetus genetically identical to that graft in her uterus.

We now know that variables like the size of the graft or its white-blood-cell content directly influence the host's ability to recognize and destroy grafts, and do so in predictable ways. But there are seemingly unpredictable differences that somehow arise in genetically identical animals. Some must be related to the experience of the animal both before and after birth, but for many models of

*At least part of this paradoxical finding can be explained on the basis of random development of immune responsiveness to rare antigens.

complex problems, genetics has not proved to be the key that unlocks the secret door to understanding. And the problems of purely genetic explanations go well beyond one discipline.

There is a general state of uncertainty about exactly what a gene does to make certain biological events happen—an uncertainty that was hardly expected in the simple origins of genetics, and one that still clouds our decisions of when, where and how to apply genetic theory to the human problems that interest us most. And this is what this book is all about.

To explore these issues, *Genetic Politics* must deal as much with psychology, politics, sociology and the theory of knowledge itself, as it does with genetics proper. Chapter 2 explores the sometimes painful psychological consequences of newly acquired genetic knowledge. An approach to understanding the limits and implications of predictions based on this genetic knowledge is set forth in Chapter 3. In Chapter 4 we will look at the consequences of the advent of mass genetic screening, and in Chapter 5 at the relative effectiveness of legislation to stem the wholesale rush to apply genetic labels and to press genetics into nonmedical areas of social policy. To fully understand genetics, *Genetic Politics* also deals with geneticists. The motives and psychodynamic forces that undergird genetic research and its applications are examined in Chapter 6.

In the final sections of the book, we will be looking at genetic politics through the glass of genetics itself—that is, by adopting a skeptical view of what genes can do. The way genes and environments work together and why their interaction has been so easily misunderstood is examined in Chapter 7. Chapter 8 explores the relevance of genetics to epistemology, or how we know what we think we know is true; Chapter 9 the error in reducing higher functions to their genetic roots; and Chapters 10 and 11 the limits of science itself as a means of knowing man and the natural world.

Finally, Chapter 12 explores the way in which humility and respect for ignorance should be part of our search for an understanding of the action of genes. In the end, the book examines why we have failed to acknowledge the ambiguity in our projections of what genes do and when we know we have reached the outer limit of what we can safely get to know. The real lesson from the genes may be that accepting complexity and ambiguity in nature is one of the preconditions for survival in an uncertain world. This book explores the realm of our genes with this ultimate question in view.

Genetics and the Sense of Self

What to the unempathic scientist is a chromosome is the heavy hand of immutable destiny to the victims: on receiving the genetic information, the patient may feel transformed into a freak, no longer fully human. Those who feel this is an exaggeration have not treated people afflicted with depression, hopelessness, or psychosis as a result of learning such a truth.
—ROBERT J. STOLLER, *Psychiatry and Genetics* (1976)

Virtually every human culture has believed in the existence of an unseen hand behind human action. The Greeks vividly imagined dark forces driving man to deeds of otherwise inexplicable passion or evil. The Babylonians believed that the birth of children with congenital malformations could foretell—and in some instances determine—the course of terrestrial events. It should surprise no one, then, that genes have become a modern-day substitute for the Furies of the Greeks, or for fate itself.

Now biologists seem to be telling us that virtually all the essential features of our development come from the genes. Is it any wonder that we look to this inner force with comparably mystical visions?

But how close those genes really are to *us*, to our innermost selves, is much less clear. And what, if anything, in our sense of self is predicated on genetic knowledge is even less clear. Author James Agee once provided an insight into the possible genetic contributions to our psychobiological makeup.

In an intuitive leap, Agee described his own self-image as if it could be reproduced by fusing the images of all of his progenitors. Agee took all the photographic negatives of his ancestors and superimposed them, one on top of the other to develop a composite image of his lineage. We can only imagine what the final picture looked like, for no actual image has survived.

It would probably be a diffuse picture, relatively clear at the points where the nuances of his ancestor's body or facial structures reinforced themselves, and blurred where there was no structural similarity. For Agee, such a composite image conveyed a powerful sense of the way he imagined genes to work.

Such a view *appears* intuitively correct, yet it is actually a distorted version of genetic reality. Belief in a blending theory to explain our inheritance was shared by great thinkers from Plato to Darwin. Indeed, for centuries, some version of a blending concept appeared the only way to reconcile the fact that an offspring always seemed to be a meld of his parents' features. Later, Agee came to believe that we are always more than our parents' genetic contributions. He fervently believed that we can transcend our inheritance—or be victims of it, but we are always something more than our genes.

All this, or rather some of this, has changed since the rediscovery in 1900 of the work of Gregor Mendel. It is no longer possible to shrug off our inheritance with such ease. We now know that inheritance is not a blending but a reordering of the basic units of heredity. Mendel's simple act of genius was to show that in pea plants, what appeared to be "blended" in the features of the offspring of a "first generation" actually was the result of the mixture of two sets of insoluble pairs of "particles" from their parents. When such individuals were crossbred, they invariably allowed the particles embedded in their makeup to show up again in the *next* generation in precisely predictable ratios, proving that whatever the hereditary units were, they were never lost through admixture or blending. Sexual reproduction in this view became a kind of reordering similar to what occurs when a pack of cards is reshuffled and new hands are dealt out to the players.

With Mendel, heredity should have fallen four-square into the realm of chance.

Unfortunately, Mendel and some of his latter day students like Sir Cyril Burt, were of a mind that made chance anathema. Both Mendel and Burt apparently suppressed chance variation to make their data correspond to arbitrary ratios or values. In part, it was this intolerance for the intrinsic ambiguities present in the genes that was the source of much of the abuse that flowed from the premature and arbitrary application of genetic principles and still contributes unnecessarily to distortions of our self-images.

In the early 1900s, geneticists eagerly applied Mendelian theory to describe the transmission of a wide range of human attributes and defects as if each attribute was "caused" by a specific gene. The hope offered by Mendelian genetics was that many of the most complex human traits could be reduced to their elemental origins in the genes. In the 1920s and 1930s, many of the human infirm were taught to consider themselves as deviant on the basis of Mendelian theory alone. On the basis of very inconclusive studies, or, more often, no studies at all, a wide range of human miseries, including alcoholism, hysteria and dementia, were labeled as recessive traits. By giving deviance a scientific base, Mendelian theory reinforced the belief that an internal flaw was behind every form of human defect.

We now know that most inheritance is highly complex in that the transmission of any given characteristic is partly the result of a broad spectrum of genetic determinants. Whole banks of genes are transferred and handed down over time from parent to child to grandchild. In this sense, it is "right" to recognize the family nose as a hallmark of common inheritance, but only if we recognize that the genes that "cause" it come from generations and generations of ancestors. And, most critically, no gene acts singly or in isolation from the environment.

The miracle of our genetic makeup is that despite this parochial lineage, genes from such widely diverse human origins as Eskimo and Aborigine can still assort and recombine harmoniously to form composite individuals who "blend" their opponent parts effortlessly during the process of embryonic development. Today it is even possible in experimental animals to mix the genetic contributions of four (or more) parents and still generate a single integrated animal. (More about this in Chapter 10.)

As we learned from Galton, in the normal world, each of us gets roughly half of our genetic makeup from one of our parents who, in turn, received a similar apportionment from their progenitors. We thus share about one quarter of our genes in common with a grandparent and one eighth in common with a great-grandparent.

Thus, Agee's subtle blending experiment is reproduced in nature by a complex pattern of chance interactions. On the grossest scale, this involves distribution of twenty-three chromosomes (the bodies that carry the genes) from each parent. Because each chromosome is part of a pair, each gene has a chance to have a different match on the companion chromosome. Such variability is common enough. Each person has about 6.7 percent of all genetic places unmatched, or "heterozygous,"* from among 60–100,000 possible structural genes. For all practical purposes, this means that each chromosome we get from our mother is always qualitatively different from the father's contribution. In fact, each of our mother's or father's gametes (egg or sperm) is so different, that we are never exactly like our brother or sister. There are 2^{23} different possible combinations—or 8,388,608 possible chromosomal pairings, and a staggering $2^{6,700}$ different gametes to choose from when a sperm or egg is produced. So you have some idea of the varieties of creation possible from just two parents!

Two more factors increase the odds against genetic redundancy immeasurably: the tendency of chromosomal segments to simply switch places with the corresponding segment of the matched chromosome next to it (called "crossing over"); and the possibility of mutation, loss or even replication of one or several of the one to two million genes that carry each individual's genetic inheritance.

The working parts of the genes are constructed from a special molecule known as deoxyribonucleic acid, or DNA. In most organisms, DNA is made like a simple two-stranded piece of twine, each strand snaking around the other to form the famous double helix. Each strand in turn is constructed of four repeating units called bases. The bases are matched between strands two by two to form rungs, much like a spiral staircase or ladder. The sides of this helical ladder are buttressed by a "backbone" of

*From the Greek *hetero* meaning "different," and *zygote* referring to the source of the genes from either the female zygote (egg) or male zygote (sperm). The antithetic prefix of *homo* signifies "same" and refers to the condition where male and female zygotes contribute the same gene. A person with sickle-cell trait is thus *heterozygous*, having received a gene for normal and a gene for sickle-cell hemoglobin; while a person with sickle-cell disease is *homozygous* for sickle-cell hemoglobin, having the same hemoglobin S gene on both maternal and paternal chromosome.

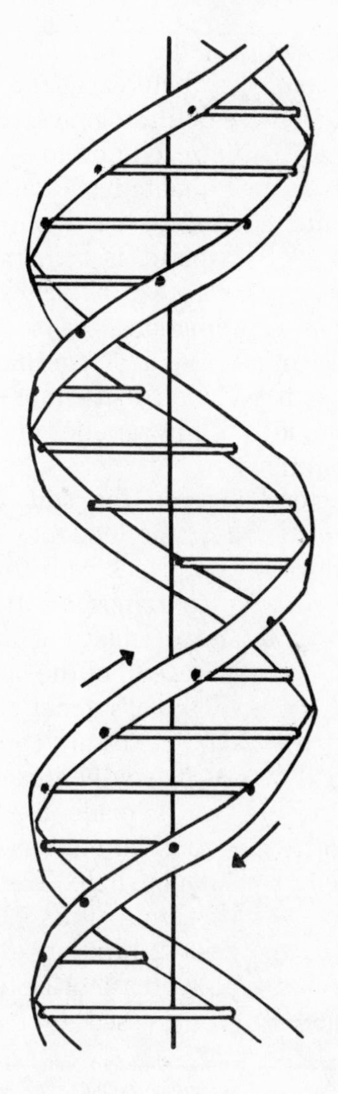

Figure 1.
The DNA Helix

Used with permission from J.D. Watson and F.H.C. Crick, from *Nature,* Vol. 171 (1953), p. 737.

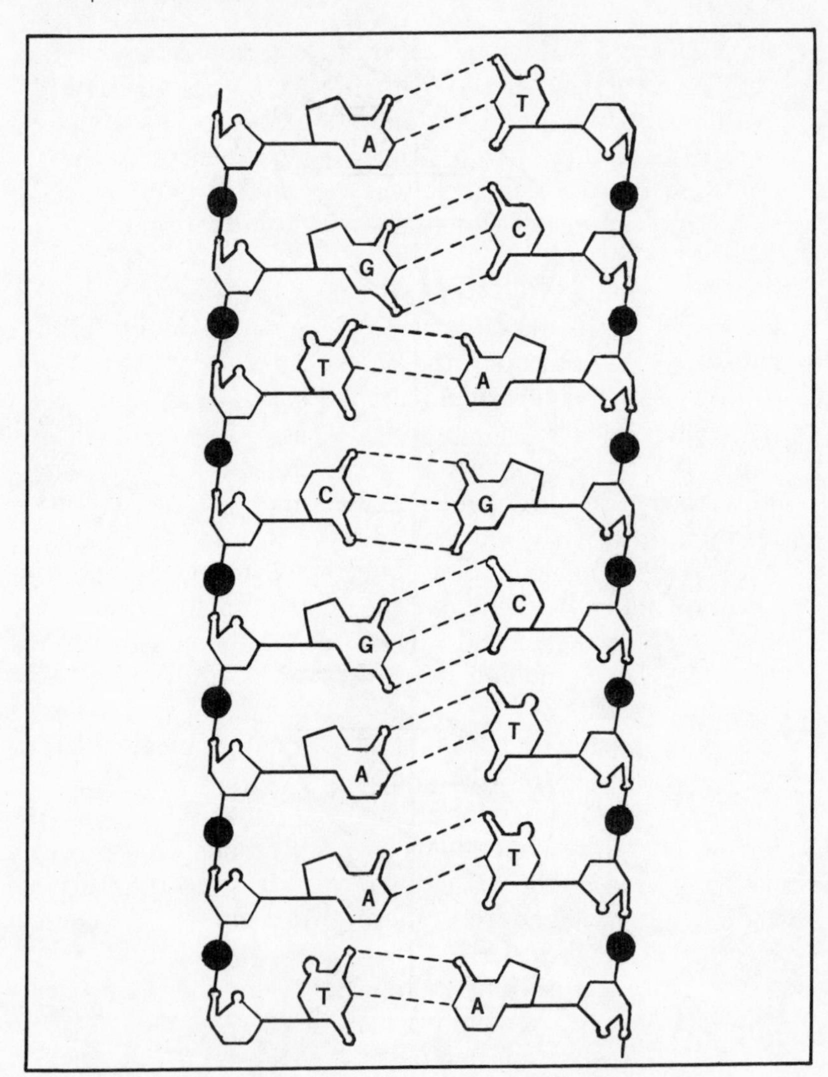

Figure 2.
The Bases of the DNA Ladder

Used with permission from Jacques Monod, *Chance and Necessity* (New York: Alfred Knopf, 1971).

phosphate and sugar molecules. Leading in one direction up the spiral ladder, the sequence of bases, Cytosine, Adenine, Guanine, and Thymidine, generate a series of random-appearing "words." Molecular geneticists use the initials of each base as letters to compose a kind of shorthand language, elementary in its grammar as CAT or TAG.

At first glance, this language reads like the gibberish produced by monkeys pecking away at a four-letter-alphabet typewriter. Yet the sequence of words—i.e., a *gene*—is usually not random at all. Instead, each group of three letters spells an amino acid—a basic protein building block. The exact sequence of words creates a "grammar" that directs the exact order of placement of the amino acids in the protein molecule. Indeed, the initials of the base pairs found along the DNA chain, Cytosine, Adenine, Guanine and Thymidine do on occasion "spell" Dick and Jane words like TAG and CAT. "CAT" in the language of the genes spells the amino acid "tyrosine." And this four-letter alphabet (G, C, A and T) is exactly the same wherever life is found. (See Figures 1 and 2 for gross and fine structures of the DNA molecule.)

These words change inexorably as the result of random forces acting over generations and generations. One word in the base sequence is altered, a letter substituted for another, or deleted entirely. These are *mutations*. CAT can change to CTT or just CT, transforming the meaning of a whole gene sequence and thereby destroying the biological effectiveness of its protein product.

One mutation, even one out of thousands, might spell the difference between life and death. For instance, a shift in the DNA message from ATC to ATT, the result of a single mutation, can be fatal if it occurs in one of the two genes that direct the construction of the alpha and beta chains of human hemoglobin, the oxygen-binding molecule in red blood cells. This seemingly innocuous shift is the first event in the genesis of sickle-cell anemia, an event that I will describe in more detail below (Chapter 3).

Variability at both chromosomal and gene level works to guarantee our individuality and assures that, with the exception of identical twins, the only time we will confront our genetic doubles is when we look in the mirror. With the exception of those persons whose immune systems are gravely impaired, no one can receive grafts of tissue from an unrelated individual without massive reaction against the insult to this genetic identity; nor, short of cloning, can he be re-created by any known process. We all face our genetic destiny alone. No amount of engineering or reproductive manipulation can change that fact; each of us is a biological island, thrown together by chance forces that ensure our uniqueness—and our individuality.

This biological bias in favor of human individuality contributes to each person's sense of self. Ultimately our genes do have much to say about who we are, in terms of race, family of origin, body type and attractiveness; and, hence, genetic knowledge speaks to something close to the core of this self—what psychologists call the *body ego*—a key element in the formation of our own special identity. For the ancient Greeks, this element was indistinguishable from the *cryptanthroparion*, or the unique, inherited destiny of each person. For us, the genes may be perceived as being similarly internal and fixed.

If this fundamentally somatic self system is felt to be "good," it can serve as the basis for high self-esteem. Seen as "bad," it can be ruinous. A flawed body ego, poor self-esteem, and environmental handicap can help to create a sense of personal impotence and predestination—one that is often frighteningly real. But when that body image is seen to be inherited, intrinsic and inextricably bound up with one's biological makeup, it can generate a sense of doom.

Take for instance the agonizing letter written by a sixteen-year-old girl to Miss Lonelyhearts—she was born without a nose:

> What did I do to deserve such a terrible bad fate? Even if I did do some bad things I didn't do any before I was a year old and I was born this way. I asked Papa and he says he doesn't know, but that maybe I did something in the other world before I was born or that maybe I was being punished for his sins. I don't believe that because he is a very nice man. Ought I commit suicide?
>
> Sincerely yours,
> Desperate

Both the girl's lament and her father's reply touch something very deep in each of us: we need to believe that, whatever the cause of our deficiencies, it is external to us—and, if we are a parent, that we cannot be responsible for our offspring's defect. Obviously, powerful psychological forces of denial and a need to explain the inexplicable color all our interpretations of life events, but we keep our biological identities at arm's length by powerful psychological forces when we suspect that self-image to be flawed.

Experience with persons who have a genetic disorder or disease suggests that genetically based explanations appear to be among the most difficult for people to accept or assimilate. Psychologist Abraham Maslow has advanced a model in another context that explains this reaction: we each have what Maslow calls a "gradient of fear of knowledge" in which the more impersonal and remote from our personal concerns that knowledge is, the more easily we can accept it. But, as the knowledge moves closer to our personal core, it becomes increasingly threatening. And genetic knowledge cuts directly to that core.

A person need not necessarily be physically affected by the gene that she carries to be psychologically disturbed by its presence. For instance, the woman who bears on her X chromosome the gene that causes the progressive weakness of muscular dystrophy in her male offspring or the devastating fragility that accompanies classic hemophilia cannot help but believe that she herself is somehow tainted or cursed. Half of all her sons will inexorably develop a degenerative condition that will preclude their achieving the traditional cultural hallmarks of manhood. In this way each child may be both a reminder that she cannot bear normal offspring and a hallmark of her own imagined biological inadequacy.

Ultimately, she may come to think of herself as totally responsible for her child's infirmity, since half of her egg cells will come from a lineage that carries an X-linked defective gene, even though she is spared their adverse effects by virtue of the process that inactivates half of the "flawed" X chromosomes in her body cells. Such guilt plagued the Romanov queens and, later, the line of descendants from Queen Victoria, who inherited the gene for hemophilia. Tsarina Alexandra believed that she was totally responsible for the weakness of her son Alexei, the heir apparent to the throne of Russia. Yet, her son's bleeding episodes were often self-induced, an act of rebellion and denial against his own flawed self. A related problem of guilt can be seen in this case study (used with permission, Albert Morascewski, Pope John XXIII Commission on Education and Genetics).

Case Study

Mrs. A. is a twenty-year-old woman who was nine weeks pregnant when she was referred to the counselor because it was suspected that she was a "carrier" of muscular dystrophy. She is one of five siblings. Her mother's fifth pregnancy ended with toxemia and convulsions, and it produced a male child, now aged ten, who has muscular dystrophy. Her mother was hypertensive, epileptic, and severely debilitated following this pregnancy. The brother experienced muscle weakness at age six, and in the following year he lost the ability to climb stairs. Duchenne-type muscular dystrophy was diagnosed, and he became bedridden within three years. Serum enzyme levels, which are elevated in about three fourths of the carriers of Duchenne muscular dystrophy, were examined for each of his four sisters at the time of diagnosis. Two of them, including the client, were diagnosed as possible carriers.

Mrs. A. was extremely depressed. She was unable to sleep or concentrate and was subject to crying spells without reason. Five months earlier she had married a man who, by a previous marriage, had a three-year-old child with some form of debilitating muscular disease. She felt that she had made a poor adjustment to her marriage and to her sick stepson. Her experience with her mother's pregnancy, which she had observed when she was ten, left her terrified of becoming pregnant. Her difficulty in adjusting to her marriage and to her seriously ill stepson added to her desire to avoid pregnancy. She feared that a second child, even if normal, would seriously strain her marriage. Her risk of producing a dystrophic child who might suffer the fate of her brother added to her fear. She had been unable to use oral contraceptives because they produced nausea and they seemed to accentuate her depressive feelings. She was using a vaginal foam contraceptive when she became pregnant.

Now that she was pregnant, she wished to determine whether her carrier status could be confirmed. If she were a carrier, she felt she would elect to have amniocentesis and abort any male fetus. She already understood that if she were a carrier, one half of her male offspring would be affected. She felt anxious to preserve the pregnancy if carrier status could be ruled out or if she could determine that she was carrying a female child.

Several serum determinations were done, and none of these was elevated, suggesting that she was not a carrier after all. On the basis of this reassuring though not definitive information, she chose not to have amniocentesis and to carry the child to term. Later in the pregnancy (twenty-four weeks) she had a severe depressive reaction and strongly regretted having passed up the opportunity to determine the sex of the infant and abort if the child were male. She felt she had made a serious mistake by assuming any risk of delivering an affected child. Mrs. A. delivered a male infant and suffered intense anxiety while still in the hospital. She was lost to follow-up very shortly. Several rapid changes of address made it impossible to learn what happened to her or to the child. The changes of address led the counselor to suspect that her marriage was breaking up.

Genetic counselors report that such anguish and denial are common among the genetically afflicted. French psychologist Jean Piaget has remarked that the closer a thought or feeling is to the self, the less likely we are to be aware of it. One might add that the less we are aware of these thoughts and feelings, the more likely we will deal with them irrationally. This may explain those circumstances in which we inject such extra meaning into the word *genetic*. Our unconscious projection of the inner metaphorical meanings of the word seems to increase directly in relation to our fear of the phenomenon that we are describing. One thinks of an incident in James Dickey's *Deliverance*, where a group of city men come across a physically deformed and sightless Appalachian boy. Their first remark—"Look at that genetic depravity!"—indicates an undercurrent of evil that will eventually overcome them. The associations to the term seem to denote Biblical genesis, original sin; they may also have something to do with the incest taboo, the crucial importance of (and the essential limitations on) relationships within families related by blood. In this case, the child's defect may well have been the product of backwoods licentiousness or inbreeding. But the fact that he was able to sound out and play a veritable fugue on his banjo confounded all expectations that physical deviance always connotes a deeper derangement. Writers like Edgar Allan Poe and movie makers like Tod Browning (see his film *Freaks*) have played on this perverse means of generating horror. Seeing normality in the face of disfigurement is almost always profoundly distressing, as if we are unable to accept the core basis for our humanness.

The tendency to react to the word *genetic* and genetic information in terms of such deep-seated psychological issues is reflected in current usage of the term.

Leon Kass has pointed out that we often inadvertently move from the possessive form to that of identity when we speak of a person with genetic problems. We stop saying so-and-so *has* a genetic defect, and we begin to speak of him as *being* that defect. Medical practitioners often describe a child with an extra Number-21 chromosome who manifests the symptoms of Down syndrome as a "Downs." And all too often we use the phrase, "Johnny is a Mongoloid" or "He is a Downs," rather than say "Johnny has Down syndrome." Declaring the child to be the disease perhaps makes it possible to see him as irrevocably different from us. This simple linguistic shift may also make it much easier for us to decide to abort a fetus with genetically related problems; after all, we are destroying a disease process, not a person-to-be.

The act of acquiring genetic knowledge can thus become both a means and an end for changing the destiny of those not yet here. Genetic knowledge is a special kind of foreknowledge that can be used to label an individual as well as to predict some of his or her life prospects. The moment one acquires a "bit" of genetic information about a fetus (for example, whether it is carrying a particular gene or chromosomal pattern), one has changed that fetus. You tell him (and sometimes others) something about where he came from and who is responsible for what he is now. You project who he may or may not become. You set certain limits on his potential. You say something about what his children will be like and whether he will be encouraged or discouraged to think of himself as a parent. In this way, the information you obtain changes both the individuals who possess it and, in turn, the future of that information itself.

Perhaps most importantly, your changed expectations may alter that person's chance to develop normally. A case in point is the labeling that was applied to boys with an XYY chromosome makeup. It was not so long ago that our society declared these children to be tainted, enough so that the prenatal detection of ambiguous chromosomal configurations was reported to parents as indicative of nonnormality in their offspring. Others regarded an XYY chromosomal constitution as inimical to the social welfare. At least one XYY fetus was aborted during this period, suggesting the pervasiveness of the public and medical perception of this "defect."*

Some recent biographical studies provide support for the contention that genetic knowledge can powerfully shape our perceptions of health and disease in our children. They are particularly interesting in terms of the enormous importance that the word *genetic* takes on when it is put

*While an extra Y chromosome contributes to an increased probability of mild mental retardation, many XYY boys are indistinguishable from their peers.

before words denoting a disease process. As Lewis Thomas has observed, we react in an almost instinctive bias when serious illness strikes. When the cause is vague, we often believe that the disease is caused from within, as if it were arranged by intrinsic dooming errors built into the system.

For instance, when photographer Edward Weston discovered he had Parkinson's disease—a progressive neurological and muscular affliction—he was sure the disease was genetically caused, a direct read-out of his flawed nature. His first reaction was anger and self-hate—feelings that were so intolerably strong that he was soon believing that others hated him for his internal weakness.

This initial reaction gave way to a fatalistic acceptance of what Weston believed to be the inevitable progression of genetic disease. But when a friend chanced to remark that Parkinson's was not genetically caused, Weston's attitude changed quite remarkably. He became able to accept his condition knowing that its cause might lie outside his body, not within it. Since the condition was not genetic, he suddenly felt that he might do something about it.

Folk balladeer Woody Guthrie went through a similarly agonizing experience—but without the final psychological release given by Weston's realization. When Woody was eighteen, his mother died of a mysterious malady that caused her mental deterioration. No one then (circa 1930) knew that his mom's condition was the dominantly inherited genetic disease then known as Huntington's chorea, a fact that ensures that each descendant has a 50 percent chance of receiving the gene and being similarly affected.

In his middle years, hallmarks of this disease appeared in Woody. He was often depressed and irritable, and his walk became lopsided. In 1952 he broke out in rages of inexplicably violent behavior. His family reluctantly hospitalized him soon after.

For years thereafter, Woody's illness was misdiagnosed;

often as not, he was labeled "schizophrenic," a process that recurs today with disturbing regularity. Only as a result of his wife, Marjorie's, constant pleading to the doctors was the correct diagnosis finally made. By then it was too late for Woody to come to terms with his illness.

Woody was unmanageably uncoordinated and, most tragically of all, unable to sing or play his guitar. Woody died in 1967 at the age of fifty-five, leaving a potentially tragic legacy to his children. Whether or not Arlo Guthrie (or, for that matter, Arlo's children) will have the disease is as yet unknown, as is the fate of Arlo's two sisters. But they live with the constant reminder of what might be their fate as the result of watching the progression of Woody's own disease.

We know about the Guthrie family because of the courage and forthrightness of Marjorie Guthrie. For her, one of the greatest stigmas of the disease was the personal sense of shame that accompanied its occurrence. "Huntington disease," as it is known today—in a partial effort to diffuse its association with the choreic movements that are typical but not universal accompaniments of its progression—is still often greeted with the abject terror of a curse. Many persons who may carry the gene for Huntington disease anticipate the advent of the uncontrollable muscular movements and dementia with a sense of doom, and even advocate sterilization of their as yet unaffected children. Others deny the possibility of their having lost out in the genetic lottery that ensures that, on the average, 50 percent of the children of an affected parent will get the disease.

Less commonly, the news that a malady is genetic may come as a relief. If "genetic" implies that nothing can be done, it can be an absolution to discover that what you have you "got" from birth—and bear no responsibility for.

Thus, Mark Vonnegut, the author of *The Eden Express*, takes solace in learning that his schizophrenia was probably caused by an internally disordered biochemistry (which he presumed to be genetically determined), not by schizophrenic mothering. The issue for Vonnegut seems to be partly assuaging responsibility—partly a sense that if there is some physical cause, there must be some physical cure.

For others, even the knowledge that one is responsible for what ostensibly appears as a virtually innocuous genetic "disease"—like albinism—can be devastating. Take, for instance, the case of albinism in non-Caucasian cultures. A Mexican-American family responded with extreme distress to the birth of two sons with a mild form of albinism. In each of the male children born in the family, this condition was manifest in having very pale skin and only barely pigmented hair. As with all "autosomal" recessive conditions (those not linked to the sex chromosomes), mother and father had to have contributed to their children's genetic makeup equally to produce an affected child. However, the father of the family was a deeply religious man who supported the traditional patriarchal values of his subculture and, hence, accepted responsibility for his sons' fate. He believed that one's children are a signature of one's manhood and was accordingly shamed by his children's lack of color, a highly regarded index of strength and virility in Latino culture.

For this father, as with many other parents, the fact that one has sired a defective child reflects on one's own sense of self. As researchers have repeatedly found, the discovery that one has passed on any genetically based disease can be deeply disturbing and can lead to profound changes in self-image to stigmatization, and even to self-alienation.

As Washington theologian John Fletcher has uncovered

in the first systematic study of the experience of amniocentesis,* discovering the genetic fate of one's children and ending a pregnancy may lead people to the brink of their understanding of their own nature—and their parental bonds. For most of the twenty-five couples he studied (all of whom had already had one genetically affected child), undertaking amniocentesis was, in fact, an experiment to get a healthy child. One of the mothers early in the study reflected on this and said, "These days you have a choice about having a healthy baby."

In his early interviews there was a buoyancy, a hope for the "healthy life," which often split the parents' loyalty to their damaged child from the new norm they imagined was open to them by amniocentesis. Almost to a person, these couples valued health and intelligence above all else. This value set had heightened their expectations for these qualities in their children. In fact, they had transferred their commitment to bearing a child with such "normal" attributes to their *own* sense of adequacy and success. Thus, pregnancy was a test. For many, another defective child would be proof positive of their flawed nature, an ultimate undoing of their sense of self.

When, as in three cases, the experiment ended in failure, all of the women elected to be sterilized. When asked why not the husband, one woman replied, "It's my fault, why should he have to pay for it." This self-imposed sense of genetic responsibility and guilt appeared to be markedly impressed on all the female members of the pregnancy partnerships. One woman receiving a positive diagnosis stated, "You spend all your life looking at pictures of pretty babies and their mothers and thinking

*Amniocentesis entails extracting a small amount of fluid and cells from the amniotic sac, usually between 14 and 16 weeks of pregnancy, for the diagnosis of possible disorders in the fetus.

that will be you. It's pretty gruesome when you are the one who is different."

Very rarely, an individual is able to use the distinguishing stigma of a gene-associated condition constructively. Rock stars Johnny and Edgar Winter both suffer from a more serious form of albinism than that described for the Latino family. In the Winters' albinism, blindness is almost always inevitable.

In Edgar's case, an acute consciousness of his uniqueness led both to his withdrawal from his peers and to his heavy investment in the private world of music. Describing his adaptation, Edgar once told a reporter, "You know the way kids naturally are if you're fat, crippled, or in any way defective. They tend to leave you out. So music became my identity and replaced the normal activities that otherwise would have filled my life."

If even a small amount of genetic deviance can lead to social isolation, what happens when formerly invisible genetic deviance is uncovered? This was one of the unrecognized pitfalls of the rush to push genetic tests onto whole populations.

Large-scale psychological profiles of populations who receive genetic information through counseling reveal a repetitive pattern of denial and repression, which again points to the power that genetic imagery still holds for us. Between 21 and 75 percent of the patients who go through even the most scrupulous genetic counseling have difficulty in remembering or assimilating the information they are told, especially when it pertains to the consequences of their own genetic makeup. As a final, albeit indirect, index of the intra- and interpersonal stress of such knowledge, one English study reported three times as many divorces among the 421 participants who voluntarily underwent genetic counseling as among a like number of controls.

Whatever the outcome, the powerful psychological impact of genetic data and information is now well established. But what are their realities? What can we actually predict from a knowledge of genetic underpinnings? The full extent of genetic variation and its implications are just now being fathomed.

3

The Predictive Power of Genetic Knowledge

Nothing is more certain in science than that godly parents beget godly children and ungodly stock spawns a godless brood.
—ALBERT EDWARD WIGGAM, *The New Decalogue of Science* (1924)

Genetic data that reveal a hitherto unimagined richness in the genetic makeup of persons is constantly being uncovered. An extraordinary amount of variation is observed, from differences in chromosome arrangement and composition (as shown by newly uncovered "banding" techniques) to variations in the amounts and kinds of biochemicals that can be found in the blood. The varieties in observed chromosomal structure are not surprising when one appreciates the variability present in the genes. Among the first genes studied in fruit flies, two or more varieties were found on as many as 30 percent of the five to six thousand available sites. Biochemical individuality is a direct result of these opportunities for variety in the gene pool. Any given individual in a species will have anywhere from 7 to 12 percent of its gene pairs in different forms (i.e., in the heterozygous condition). In humans, for instance, this means that a significant number of persons may have the genes for AB blood type instead of pure A or pure B. When more than 2 percent of the population is heterozygous for a given pair of genes, that variant is called a *polymorphism* (from the Greek for "many forms").

The existence of such an extraordinary amount of diversity creates at least two conceptual problems: first, to ascertain the evolutionary significance of genetic variety; and second—and more important for the enterprise of this book—to find the meaning of such diversity in terms of the quality of human beings. Is the fundamental essence of the human form flawed by this diversity as pessimistic observers like John Milton* have imagined, or is it a signal of heretofore unappreciated variety? Evolutionary biologists themselves are torn between two schools of thought. Motoo Kimura, of Japan's National Institute of Genetics, believes that this newly observed genetic variation is, for all practical purposes, a neutral phenomenon, reflecting the random acquisition of new genes through chance mutation, with little or no adaptive significance. In this view, the biochemical differences that we observe among persons is largely the reflection of genetic "noise" and of no real significance for their welfare. But much of this variation is too common to be explained on the basis of random mutations alone.

The contrasting view is championed by geneticists like Francisco Ayala of the University of California at Davis, who believe that most variation that we see does have special significance. In the past, some geneticists, notably Hermann Muller, the discoverer of radiation-induced mu-

* Milton wrote in *Paradise Lost:* "Man by number is to manifest, His single imperfection, and beget, Like of his like, his image multiplied, In unity defective, which requires Collateral love and dearest amity."

tations, postulated that most of this diversity, in fact, constituted a genetic "load" on the population. Other geneticists, unlike Professor Kimura, hold a selectionist viewpoint, which regards the unusually high prevalance of some polymorphisms, like that for hemoglobins (e.g., sickle-cell trait) as signaling some kind of advantage to the carrier. Even the most chemically inconspicuous genes, in this view, can have powerful biological implications. A gene that subtly modifies an organism's ability to respond to starvation or to deal with external pollutants, for instance, can have far-reaching effects on the carrier's prospects for survival. It is these last possibilities that deserve the closest scrutiny for us. Diabetes is a case in point. Recent experimental evidence in mice suggests that carriers of one of the recessive genes which gives rise to diabetes in homozygotes actually endure the privations of starvation better than mice with the normal genetic makeup. Although human diabetes is almost certainly more complex, such evidence gives credibility to the hypothesis that the "carrier" diabetic state is somehow a *thrifty* genotype that confers a selective advantage on the person in times of duress. Diabetes, the result of the interbreeding of such carrier individuals, can then become more prevalent, even though the affected individuals have reduced reproductive capabilities.

The critical test of the quality of genetic variation is made by following the results of consanguineous marriages. If most of the recessive genes that are carried in a given population are deleterious, then inbreeding will reveal their presence in a higher incidence of fetal or newborn deaths, or indirectly through diminished fitness of adults. Such disclosure occurs because lethal autosomal recessive conditions produce their fatal effects only in individuals who receive two copies of the defective gene, an eventuality greatly enhanced by the childbearing by persons with a common ancestry—and hence a common opportunity to have received the same recessive gene.

Certain rural villages and cities in Japan, because of their traditional isolation from commerce and the outside world, are places where (as in Appalachia) there is a considerable degree of intermarriage—first cousins marry first cousins more frequently than usual, uncles marry their nieces, and so on. Several such communities have provided the testing ground for what was in the early 1950s a dangerous and provocative question: Just how "genetically healthy" are human beings as a species?

Each consanguineous marriage has a probability of bringing together lethal or deleterious genes that are otherwise hidden in their recessive state; for instance, a first-cousin marriage brings together one eighth of the total recessive genes carried by both individuals. Astonishingly, instead of the massive die-off predicted by believers in the genetic-load theory, the Japanese subjects got a relatively clean bill of health.* Historical cues to this outcome were there to be seen. The Eighteenth Dynasty Ptolemies of Egypt, for example, were purported to have practiced generations of consanguineous matings with impunity. Of the thousands of possible deleterious genes, the number of actually lethal ones carried by each person was found to average between only *three* and *five*, with appreciably more genes being only slightly deleterious. Cavalli-Sforza and Bodmer, in their *The Genetics of Human Populations*, summarize the more recent view by declaring, "Almost everyone carries the equivalent of more than one lethal recessive gene. . . ." Genes which are deleterious in a double dose (homozygotes) *and* unusually prevalent are uncommon. Examples, like hemoglobin S,

* In this study infant mortality in first-cousin marriages went from a normal 1.5 percent to 4.6 percent.

can almost always be shown to have conferred some selective advantage on carriers in the past. Strongly deleterious recessive variants (i.e., lethals), by comparison, are almost always a "burden" in the sense that they undercut the adaptive abilities of the carrier. Thus, recessive lethals in fruit flies reduce the overall fitness of their carriers by some 7 percent; and the gene for ataxia telangiectiasia, which is lethal to humans who receive a double dose, reduces the probable fitness of its carrier by a comparable amount by increasing the risk of fatal malignancies at an early age. Thus, while no one is genetically perfect, most of the observed variations are either neutral or of some slight selective advantage.*

If genetic variability is thus measurable, and a given discernible difference counts in terms of our adaptations to our environment or our behavioral idiosyncrasies, then we have an immense new power in terms of what we can say about the prospects for the quality of life of any given person in any given environment. More ominously, we can also begin to measure the social cost of a person, for instance, those whose makeup from birth registers the likelihood of specific medical difficulties.

Initially, only a minuscule portion of the kind of data we acquire will be of any value to the person himself. However, much of it will be potentially valuable to health

*Distinction should be made between selective and adaptive advantage. A gene which confers a selective advantage on its carrier increases overall fitness by giving an individual a reproductive edge over noncarrier individuals in the same population. Hemoglobin S in malarial environments is such a gene. Adaptive advantage is a relative term which describes only the physiological wellness of an individual in certain environments. Selective and adaptive advantage, while usually concurrent, are sometimes independent. Thus, the hemoglobin S gene seems to be selectively neutral in normal environments, but adaptively disadvantageous when carriers are exposed to microenvironments (e.g., low oxygenation under anesthesia). Similarly, carriers for the dominant gene for Huntington disease characteristically leave behind substantially more descendants than do noncarriers (for as yet unknown reasons), yet are at severe adaptive disadvantage as they reach the age when their disease process appears.

planners, and other portions will be seen by insurers and employers as critical data for *their* economic investments. The cutting edge of the value of all genetic information thus rests with its predictive weight—how much does knowing something about the genetic makeup of a person tell us about his prospects for well-being?

At present, almost all our prognostications are clouded by the uncertainty of our knowledge about genetic causation. With the relatively rare exception of those genes that are known to be associated with metabolic or other clinically apparent genetic diseases, *we simply don't know how tightly any given gene ties us to our destiny*. In those areas where we are beginning to make inroads, some of our knowledge appears firm, while elsewhere our knowledge is statistical at best.

Uncovering the molecular basis for sickle-cell anemia is an instance of one of our triumphs of specificity. Here the discovery of the genetic basis for the molecular misconstruction of one of the 140-or-so different varieties of hemoglobin, unlocked part of the secret to the disease process itself.

There are four chains of building blocks known as amino acids in hemoglobin, and minor changes or substitutions in any one can be inconsequential or catastrophic, depending on the area affected. In sickle-cell anemia the code change that leads to simple replacement of one amino acid by another—a glutamic acid with a valine—in one of these chains leads to a complete restructuring of the hemoglobin molecule. When the amount of oxygen available in the blood diminishes, these hemoglobin molecules lose their oxygen and stack up like crystals, causing their encasing red-blood-cell membrane to collapse into a characteristic *sickle* shape. And these sharp-edged cells pile together and jam up small capillaries, stanching the flow of blood-borne oxygen to vital tissues, particularly in the joints and spleen. In this way, knowing the underlying genetics of a condition pointed both to its origin and the reason for its health consequences.

But, at other times, the presence of a genetic variant allows us to say only that the *possibilities* of a deviation from health are shifted toward the occurrence of a particular deleterious event. The archetype of this kind of predictive situation is to be found in the genetic system that provides us with our biological identity.

Each of us has his biological uniqueness encoded in a complex set of genes. These gene complexes are part of a "tissue compatibility" system called *HLA*, which encompasses a region on a single chromosome (No. 6) where there are four distinct clusters of genes known as the A, B, C and D loci. The genes in this HLA system are expressed by different chemical constituents known as *glycoproteins* on the surfaces of our cells. Until recently, the real purpose for having so many glycoprotein "markers" on each cell was but vaguely understood. Was it to protect us against potentially disease-causing invaders by helping our immune systems separate self from nonself? Or was it part of a surveillance system that allowed us to patrol our own body cells to efficiently seize and destroy those cells that might deviate toward malignancy? Many immunologists now think that the major function of these chemical markers is geared to specific capabilities of our immune system. Some researchers believe that this system ensures our biological individuality by perpetuating genetic variety. (One way to encourage variety is to have the HLA system tied to beneficial immune reactions against the fetus. In this way, on the average, genetically *different* fetuses in the womb would do better than do those genetically similar to their mothers.)

Whatever the true reason for the existence of the HLA

system, recent discoveries relating the presence of these markers to susceptibility to disease states has markedly shifted the focus of our attention.

The genes that find their expression in the HLA system number in the hundreds. Genes in at least three components of the HLA system (loci A, B and D) have been linked to heightened susceptibility to as many as forty different disease states, many of which include self-destructive immunological reactions known as autoimmune diseases. For instance, the gene that codes for the antigen or cell-surface constituent known as HLA B8 is found with increased frequency among persons with self-destructive immune reactions and is thought to be associated with diabetes. Another antigenic determinant, DRw4, occurs commonly in persons who develop severe rheumatoid arthritis.

A surprising portion (up to one fourth) of seemingly healthy individuals with another variant, B27, turn out to have an extraordinary incidence of latent back problems. For up to 95 percent of the males who have the debilitating back disorder known as *ankylosing spondylitis,* the B27 gene provides a diagnostic confirmation that their condition is a gene-linked variant of this disease. The fact that ankylosing spondylitis is common (about three million persons in the United States alone have it), painful, occasionally debilitating, and resistant to treatment makes this knowledge weighty indeed.

This news is hard enough to take. But couple the knowledge that this marker is transmissible—not by the ordinary means of contagion, but through the germ plasm—and you have a scenario resembling an ancient Greek play. The person with the HLA B27 gene will pass it on to half of his offspring. While having the gene itself does not spell doom to the individual, since it does not *invariably* lead to later disability, the risk of chronic back disease or more severe autoimmune processes nonetheless colors the prospects of quality of life of any newborn.

The risk that any person with the B27 gene will develop one or more of the various disease states associated with the gene is still being assessed. The relative-risk figures that are known, however, are not comforting. The person with the B27 marker is between 40 and 110 times more likely to develop ankylosing spondylitis than is a person who does not bear the responsible gene. While the gene is relatively uncommon in most populations (e.g., 8 percent of the white American population), it is remarkably common in others. Approximately 50 percent of the North American Indian population carry the gene, and as one would expect, many of them suffer from the constellation of disease states linked to B27. Couple the fact that some experts have already counseled B27-affected individuals and their prospective employers against employment in occupations which stress the lower back, with a racial (e.g., American Indian) predilection for high B27 frequencies, and the potential for using genetic differences to justify discrimination is readily apparent.

Other genes in this same system appear to determine biological attributes other than disease susceptibility. The gene product known as Dr7 for instance, is found in people who don't respond to influenza vaccines, and others are linked to a reduced risk of diabetes. These gene associations are bound to grow in number. Their clinical implications are obvious. Less clear are the implications for public policy—When and how (if at all) should we institute screening programs to detect these genes? Should that screening net be extended into the prenatal period? Who should have access to the subsequent information that these programs will generate? And finally, what will we know when we have amassed this new data base that we did not know (and could not have known) before?

What if a test became available to assign each person a risk estimate for their propensity to develop painful, debilitating and, ultimately, costly diseases like those associated with the HLA series of genes? The question is no mere thought exercise. Prenatal tests for HLA antigens are eminently feasible, since these substances are among the first to be expressed during mammalian development. Why not apply the genetic screen prior to birth and thereby avoid the irrevocable prospect of a disease fraught with suffering? Prenatal tests for diseases like hemophilia, muscular dystrophy and Tay-Sachs disease are already applied this way; why not these other conditions?

One argument mustered against this position is that, unlike HLA diseases, gene-associated disorders like hemophilia and muscular dystrophy are unequivocably linked to a given gene; here the prognosis made by the genetic test is virtually absolute. We can now know, in a way only imagined before, just who will develop these conditions, and give parents the prospect of their lives with relative—but only relative—certainty. A decision to terminate such a pregnancy would then be an informed choice based on judgments drawn from fairly hard facts.

But, with the new wave of genetic information developed through HLA testing, our knowledge will be less absolute and secure. We will have only statistical associations with a life prospect. Not everyone with HLA B27 develops the diseases associated with the gene. It is unlikely that we will ever develop the certainty of prediction that we would need to make life-and-death decisions—but we will have some new data on which to plot an individual's prospects. What we should do with that information and what it means to the average person have been but poorly examined.

Firmer answers to the actual relationships of genes to disease processes will undoubtedly give way to more sophisticated interventions or preventive steps. For instance, the arthritic diseases linked to B27 may be triggered by an immune reaction to an infectious organism. Special procedures to protect at-risk individuals from exposure to such organisms might thus be warranted. But what do we do in the interim? As more and more genes are found that show these links to incurable disease states, we will be tempted to use the information we have in hand to shape our policy decisions.

In time, virtually every disease or disorder will almost certainly be found to have a genetic underpinning. Some diseases which we have presumed to be environmental in origin might well be found to have a genetic analog that closely mimics the symptoms of the disease in question. For instance, we already know that some forms of manic depression, in which a person's moods swing wildly between elation and despair, follow a pattern of inheritance that strongly suggests linkage to a gene on the X chromosome. Mothers appear to pass this psychological disorder on to their sons in the manner of classic X-linked diseases. And "unipolar" depression has at least three genetic forms.

But do such discoveries mean that in some way all of our feelings or moods are indirect projections of our heredity? Will emotions that we have always considered to flow from our innermost selves—our joys and sorrows, elation and suffering—likewise fall to a genetic explanation? How much of our fundamental human attributes will turn out to be preconditioned by genetic messages embedded deep within us?

Perhaps a glimpse toward an answer comes from looking more closely at disorders like manic depression. Most certainly, not all forms of this illness are genetic, and knowing which forms are linked to other causes has proved invaluable in designing therapeutic regimens. Classic bipolar manic depression is responsive to lithium;

the genetic form may not be. But the question of genetic underpinnings to other disease states remains unanswered—and tantalizing in the power it appears to offer.

Other diseases show how a knowledge of genetics permits a more enlightened approach to diagnosis and care. Recent studies of multiple sclerosis, for example, have shown that the same symptoms may represent two genetically different disease processes, with two different rates of progression. Certain people are turning out to have an enhanced tendency to develop multiple sclerosis and can expect to develop it in a more serious form when it does occur.

Nor is multiple sclerosis the only disease that may turn out to be strongly predetermined. Certain forms of cancer are also going the same way. The monamine oxidase content of schizophrenics' blood platelets has recently been found to have possible predictive value, so that one may be able to tell which children of schizophrenic parents will themselves develop the syndrome. Even heart disease, the number-one killer in developed countries, may fall to the genetic arrow. Recent studies show that three genes may point to as many as 20 percent of the persons less than sixty years of age who are at risk for a myocardial infarction (heart attack).

In all of this, diseases that were believed for centuries to be the result of the environment—something any of us could "catch" like a cold—may well turn out to have been inevitable for some people from the beginning. We already know that this is true for those unfortunate few who receive a double dose of the genes for Fanconi's anemia, ataxia telangiectasia or xeroderma pigmentosum—all disease states that carry virtually a hundred-percent probability of cancer for their victims. Less well appreciated is the possibility that the much more prevalent carrier of the same genes may also be at increased risk of cancer. The roughly one in 100 persons who carry the gene for ataxia telangiectasia are five times more likely to develop cancer than are noncarrier controls.

In the not-too-distant past, discoveries such as these would have been welcomed as an unmixed blessing. Humanism and Western science have traditionally taught that knowing is, in and of itself, a valuable thing; moreover, common sense tells us that when we know the source of a problem we are in a better position to begin solving it. In light of recent discoveries and applications of genetics, however, one must reexamine these truisms about knowledge.

For example, what do we do with that genetic knowledge that tells us who carries a particular gene, but not what that gene will mean for their future well-being? Or what do we do with the genes that, like the HLA series, are presently ambiguous in exactly what they "fix" or determine? Undoubtedly, increasing numbers of individuals will be placed into a genetic twilight zone where what we can tell them is colored by the uncertainty of our understanding.

If this ambiguous state of affairs were recognized and accepted by all involved, genetic knowledge would not pose the problems it now generates. As a culture, we have been unwilling to stand still in the face of uncertainty. We tend to want scientists to say more about a particular causal link than they themselves believe is warranted by the facts.

One of the manifestations of this need to know comes in the drive to devise predictive tests for latent genetic disorders. For many persons with family histories of Huntington disease, the uncertainty of not knowing in advance who among a given population of at-risk persons might succumb to the depredations of this particularly horrible disease process is almost unbearable. In the words of science writer Richard Saltus, "A sword of Damocles hangs over thousands of Americans who may carry the gene for Huntington disease."

As we have seen in the case of Woody Guthrie, the one in 15,000 to 20,000 who develop Huntington disease experiences a very rapid intellectual and emotional deterioration after an indeterminate latency period. Although the disease ordinarily makes its first appearance in victims between thirty-five and forty-five years of age, it has struck children as young as five, and adults as old as seventy.

As reflected in the fact that some 5–6 percent of victims commit suicide, Huntington has enormous emotional and dollar costs for its victims and those caring for them. Recently, a research group headed by Dr. Harold Klawans of the University of Chicago devised a preemptive test for the disease. Noting that the neural transmitter precursor L-dopa could aggravate the symptoms of Huntington, and that the disease is transmitted genetically, Dr. Klawans proposed that relatives of Huntington victims take doses of L-dopa to diagnose their susceptibility to the disease. If these participants developed symptoms under L-dopa, then one might hypothesize that they would eventually develop Huntington sometime later in their lives.

In a 1972 article in the *New England Journal of Medicine,** Klawans reported the results of a study of twenty-eight relatives of persons with Huntington. Thirty-six percent of this group, but none of the controls (who were unrelated to anyone with the disease), developed one or more signs of the disease. Several of the participants who reported chorealike symptoms—which they took to be a preview of their fate—later experienced suicidal reac-

*H. L. Klawans et al., "Use of L-Dopa in the Detection of Presymptomatic Huntington's Chorea [sic]," *New England Journal of Medicine,* Vol. 286 (1972), 1332–36.

tions. Other participants welcomed what they took to be the "clean bill of health" demonstrated by the test.

In fact, however, neither group could know whether they had "passed" or "failed" the test. Many more years would be required to prove that L-dopa–evoked symptoms had any relation to the later development of Huntington. Moreover, when Klawans realized that precipitating the symptomology of Huntington might actually accelerate the onset of the disease among those participants destined to express it, the testing program was halted. (A variant of this preemptive test, which looks for early muscle tremors, has recently been reported by a research team from Georgia.)

Among the problems that Klawans' group considered was the impact of such knowledge on the presumptive beneficiaries. Certainly, it appeared enormously desirable to be able to tell someone who faced the prospect of Huntington disease that they were, in fact, not at risk. But what about the information that would have to be given to the carriers whose doomed status was revealed by his testing? (The Georgia group is so concerned about the impact of such knowledge, that they have taken a vow of silence until they can be sure of the findings.)

Psychiatrist Lissy Jarvik asked these and other questions rhetorically at a symposium of the American Psychiatric Association held in 1974 in Hawaii. Professor Jarvik said, "Once we are able to detect asymptomatic carriers, should we tell them? All of them? Some of them?" She asked whether the relief that would be conferred on those who were found to be free of the gene was worth the certainty of doom that their brothers and sisters would simultaneously experience. And then she asked the ultimate question, "Can we hide from another human being information that may be vital in his choice of the way he lives?"

Were this question limited to what an individual might do with the information gathered, the issue might not be so problematic. But other entities in society—insurers, corporation employers and family members—could also be shown to have a legitimate "right to know." How would an individual handle these situations, and would he have control over the information? Others might, in fact, use such highly sensitive information in ways inimical to the well-being of the individual.

For example, both businesses and primary-care physicians are interested in gene-associated conditions in which there is an increased risk of heart disease. These disturbances can often be detected by looking for raised levels of lipids or triglycerides in the bloodstream. In one such condition, called type II hyperbetalipoproteinemia, the presence of a single gene confers on the affected person an almost even chance of being struck by a life-threatening heart attack before the age of fifty. Between four and eight newborn infants in every one thousand can be found with the hallmarks of this condition.

New dietary regimens *may* offer some prospect of reducing these risks, but prospective testing of the therapy will require at least twenty years to complete. In the meantime, researchers have warned that any person, particularly a male, who bears this gene will likely be denied certain high-paying executive positions.

If Klawans' or other preemptive testing of adults raises serious ethical questions about the therapeutic effects of knowing, these effects become the more critical in considering the effects of genetic knowledge on newborns. Currently, genetic research is commonly carried out in hospital nurseries, uncovering new knowledge about biological variation among infants. As we may have expected, an extraordinary amount of such variation is observed.

Some of these differences seem to be of academic interest only, reflecting normal variations of no apparent clinical significance. Other variations, however, may be indicative of impending metabolic difficulties or long-range behavioral abnormalities.

The burden of such information may far exceed any benefits to the people most directly involved in this research—the babies and their parents. Most disease processes that can be identified by genetic analysis cannot be meaningfully treated today, and profound ethical questions are associated with the development of effective therapies for the future. For instance, when unusually high levels of histidine, a normally occurring amino acid, were found in the cord blood of some newborns, researchers hypothesized—by analogy to PKU—that the infants would later develop some form of mental retardation. Given this educated guess, researchers had to grapple with the dilemma of whether to intervene or not. If they left the infants alone and their hunch about the relationship between histidine levels and retardation proved out, the infants might needlessly suffer impairment. On the other hand, if they put the infants on histidine-free diets, they could never be sure that genetically caused high histidine levels *were* related to retardation.

Beyond this research dilemma, the simple act of gathering genetic information about a fetus or newborn automatically sets in motion a chain of events that will determine that fetus' future as a person. We may discover, for example, that the fetus is a male, and that his cells lack a single critical enzyme necessary for synthesis of DNA precursors known as purines. If this is the case, we also know that the fetus picked up the critical gene from his mother, because the gene appears only on the X chromosome contributed to the male fetus maternally. Without this enzyme, the child will develop Lesch-Nyhan syndrome, and almost certainly will be retarded and subject to bouts of self-mutilation.

We may also be able to predict, with varying degrees of accuracy, the person the fetus will become in the future. For less severe gene-related afflictions such as ankylosing spondylitis, the prophecy may simply be that the fetus will not do well in a job wherein he or she has to sit for more than six hours a day.

Current genetic pretesting is but the opening wedge of a movement to predict disease states years in advance of the onset of symptoms. Such prescience will give the genetic scientist unheard-of powers. From a biochemical analysis of blood, physicans or researchers can learn things about people that in a very real sense will enable the scientist to maintain an uncanny measure of control over their lives—in terms of what work they may be suited to do, whom they may fruitfully marry or not marry, even how long they may live. In a sense, this power parallels that of the traditional shamans in premodern cultures. Neonatal and prenatal testing is twentieth-century industrialized society's version of that ancient belief that possession of any part of a person—a fingernail, a lock of hair—confers power over that person's destiny. Today, the nail and hair have been replaced by slides of fetal cells and amniotic fluid. A critical question that arises in regard to this issue is whom shall the modern shamans serve?

One answer is clearly those with a stake in identifying and excluding persons with genetically based maladies. Insurance companies are requesting increasingly sophisticated kinds of data, particularly before they will underwrite policies for newborn infants. In time, insurance companies may even request that data be obtained before birth to speed up underwriting and marketing procedures. Many researchers and physicians still believe genetic tests are a good way to define roughly the early prospects of an individual's health. That is, surely, a reasonable hypothesis. But the business sector may see the data generated by such tests as a means to identify and discard high-risk persons, even where the data are incomplete and tentative.

While genetic data can generate fear and self-fulfilling anxieties, they can also lull with false assurances. Just as uncovering a deficient enzyme system for a critical metabolic function seems to be a fail-safe indicator of subsequent sickness or death, a "normal" sex-chromosome complement seems to say that development will drive the fetus toward a female or male morphology. Likewise, "passing" a battery of tests for metabolic disorders implies that the fetus will have the prospect of accomplishing the biochemical reactions necessary for health and well-being. And, as we have seen, the presence or absence of key transplantation markers seems to provide a way to determine whether the child has a reasonable prospect of long-term well-being.

All this is almost, but not quite, true. While vast amounts of new data are available through the biochemical window of genetic analysis, projections of what the data *mean* are still most elusive. "Normal" prenatal test results provide only a means of excluding the possibility of major developmental deviations. They do not establish normality. For instance, the presence of a normal XX chromosome makeup and normal karyotype virtually rules out Down syndrome, but does not establish that the fetus within the amniotic sac will become a functional female person. A rare XX fetus develops into a biological male—and vice versa. And experts in sex determination, like John Money of Johns Hopkins University, are quick to point out that chromosomal sex is but one step toward psychocultural sexuality and the assumption of "normal" sex roles.

On another level, the air of certainty surrounding

genetic research may be used to clothe data of a far less certain and disinterested kind. The argument that genetics determines intelligence is a case in point. IQ was at one time a construct, a convenient way of predicting which students and soldiers might fail training in basic skills. It was assumed that such failures resulted from intellectual deficiencies (one of several hypotheses that might explain the failures) and that IQ tests actually measured intelligence rather than some other human characteristic (such as social background) that might also explain the phenomenon. Later studies seemed to demonstrate that those related by blood or by race or by social class were more likely to score within the same IQ range than persons unrelated by these factors.

In fact, however, there are several problems with asserting that such data prove that intelligence is genetically determined, or that it varies according to race or other factors. Weaknesses in the research designs and statistical analyses of social-science research make it extraordinarily difficult to approximate the associations between intelligence and genetic makeup. We certainly know of no gene, or even combination of genes, that "determines" intellectual power.

On an individual level, other researchers have tried to establish associations between IQ and genetics by studying identical twins adopted into different households. Yet it is generally impossible to rule out environmental effects in these studies, since twins are, as a rule, placed in highly similar settings. The possibility of showing that *group* differences in intelligence are genetically based is even more difficult. Where all factors are controlled, factors in the environment appear to largely account for lingering white-black IQ differentials. The absolute degree of genetic mixture between whites and blacks in this country cannot be ascertained, much less controlled, for the origins and genetic composition of both groups are heterogeneous.

Despite the virtually insurmountable and self-admitted difficulties involved in such research, particularly in the accurate partitioning of genetic and environmental interactions, psychologist Arthur Jensen once encouraged the use of gross black-white IQ differences to explain the "failure" of compensatory education programs. Similarly, Richard Herrnstein has used differences in IQ scores between those at the top and those at the bottom of our society to support his theory that the most intelligent persons will inevitably rise to form a "meritocratic" class. Such beliefs are asserted as genetic truths.

Such truths turn out to be limited approximations at best, since they usually rely on a statistical estimate derived for diverse populations under less than controlled conditions. The statistic most commonly used is a *heritability* estimate. In its simplest form, heritability describes the proportion of the variation observed in a given trait (like IQ scores) that can be attributed to genetic differences among the individuals tested.

The premise of this statistic is that a trait that shows appreciable variation and a high heritability will likely reflect significant genetic differences among those tested. But caution is often needed in the interpretation of results. For instance, corn plants that grow in a perfectly plowed and fertilized field and show dramatic differences in height will almost certainly have a high heritability for that trait. The plants would also seem to be demonstrably different genetically, since all measurable environmental factors that might have contributed to the differences appear to have been accounted for. The researcher might even feel safe in extrapolating these results to suggest that the observed differences reflect bona fide genetic differences among his seeds. But if even a single factor, such as the level of trace elements like zinc or copper, in the environment varied nonrandomly across the field, the same variability in height could be observed with genetically *identical* plants whose true heritability would have to be zero! Conversely, genuine genetic varieties of corn grown in a field under absolutely uniform but limiting conditions of nutrition throughout, can show minimal variability in height; yet, they would still have to be given a high heritability score, since virtually all of the observed difference would by necessity be due to genetic variation.

The limitations of heritability scores for estimating genetic differences between two groups whose norms differ, as do black and white IQ scores, is shown by using this last example. Suppose a different handful of seeds, taken from the same bag as those grown in the nutrient-deficient soil, are now sown in ideal soil. Now we will have two groups of plants with different heights, *each* with high heritability scores. Because the environmental conditions were uniform for each group, virtually all of the observed differences in height would again be due to genetic variation. But we would be wrong in concluding on the basis of the heritability scores that the two groups of plants differed significantly in their genetic makeup, since all of the difference in their average height was environmental. By analogy, we might be making the very same mistake anytime we tried to use the existence of a high heritability where two groups differed in a certain trait as implying real genetic differences between those groups.

To further compound the problem, it is now widely accepted by statisticians that several confounding variables limit the validity of heritability estimates of human IQ scores. As we have seen, a heritability score (which is the first step in extrapolating genetic differences between and within groups) must be obtained under rigorously controlled conditions. Accurate computation of a heritability figure for a given trait is impossible also where variations in the environment are not completely random. Even if one succeeds in keeping environmental variables controlled for one group in question, unless the same environmental variables are present for the other group, cross-comparisons (called between-group heritability) are of questionable reliability. Such a situation obviously obtains for most black-white comparisons, except where cross-fostering brings representatives of each group into similar environments.

In the final analysis then, at least three conditions preclude a test of the genetic hypothesis for white-black IQ differentials: (1) IQ scores themselves may be of questionable validity as measures of intelligence; (2) heritability figures may not be strictly comparable between groups; and (3) accurate measures of gene-admixture between whites and blacks needed to produce expected intermediate populations may be impossible to obtain.

The first of these three conditions has been raised on both philosophic and statistical grounds. Some philosophers argue that IQ is still a flawed construct of pure intelligence. Because an IQ score is also only an indirect measure of intelligence, it fails another test of heritability equations—namely, that the trait in question be *directly* measured.

Secondly, the presently low heritability estimate generally reported for IQ scores among blacks contrasts markedly with the high estimates found among whites, suggesting to many researchers the continued presence of environmental variables which skew black performance.

Finally, despite the discovery of increasing numbers of genetic loci that purportedly serve to distinguish black populations from white ones, such genetic differences still fall far short of the necessary specificity and completeness needed to grade each individual's white-black genetic makeup, and hence to assign a projected value to his IQ score. Since cultural factors and attitudes often go hand in glove with many of these linked genes, most notably those for skin color, it is highly unlikely that a completely culture-free genetic comparison of the two groups will ever be possible. In the meantime, adoption studies, most notably those of Sandra Scarr and her colleagues at the University of Minnesota, in which underprivileged blacks are placed in affluent white homes, continue to demonstrate the powerful effects of environment on raising the IQ of otherwise "genetically unintelligent" children.

Nevertheless, as long as beliefs of a gene-IQ link are strongly held, they may, in fact, become self-fulfilling prophecies, as those who score poorly on the tests are relegated to impoverished educational settings; and more ominously, to those home and physical environments that quashed their potential in the first place.

While it is true that early calculations of heritability provided efficient models for predicting the success of animal breeding experiments, we now know that the various measures of heritability are limited in their applicability to the human situation. Each such estimate is limited to the place and general environmental conditions that obtain at the time of its determination. And, since the environments of the groups that are most often compared (i.e., adopted black and white twin pairs) are often radically dissimilar in the real world, calculations based on the experience of researchers with adopted families, or even interracial adoptees, are bound to give skewed results.

Few, if any, genetic researchers bother to explain the distortions that result if the environments of their study are not exactly replicated by the world to which they wish to generalize. The calculations which worked so well for predicting the success of breeding for hog-back thickness, egg number, or beef girth have very little relevance to our own society. These calculations worked to allow us to breed "superior" types precisely because the conditions of the barnyard could be kept so uniform and dissimilar to the heterogeneity of human environments.

At least for now, we may hope that the heritability figures that are presented in IQ/heredity studies are not being presented to predict the success of human breeding experiments, nor to generate "facts" about differences between groups whose respective life experiences are simply incomparable.

At the extreme, we can recognize the moral unacceptability of sacrificing extraordinary numbers of individuals or regulating their reproduction to achieve a uniform degree of success in raising the intelligence of humans as was done to raise the number of eggs that a chicken might lay. And even the most "reasonable" breeding programs for the "selection" of the brightest would necessarily involve the most repressive measures and an elitism (in sperm selection, for instance) that runs completely counter to our political principles.

More reasonably, we will likely find that our calculations for the application of simple genetic principles to humans will work much less well for the complex phenomena that we want to study than they did for the simpler world of the field and barnyard. The lesson appears self-evident: Unless the limitations of a strictly genetic approach are learned concurrently with its strengths, it is likely that error, ignorance and abuse will taint the application of genetic knowledge to the solving of human problems.

What is really at stake with genetics today is the possibility of dictating the conditions for employment or education, or the predilections toward disease for whole groups based on genetic data. As long as one is prepared to read the speculative and overgeneralized conclusions of genetic analysis critically, such information may be relatively harmless. As long as one is prepared to hear genetic predictions about the destiny of groups of persons with an allowance for the limitations of genetic knowledge, such predictions may not become self-fulfilling prophecies. But ultimately, even such skepticism may not prepare us for that species of genetic knowledge that will be presented as more certain and well-tested. We need to wonder whether we are ready on a mass scale for those tests that point to a foreknowledge of our biological and psychological destinies—and for the limitation of freedom that such knowledge implies. Mass screening for genetic diseases provided the first testing ground for the impact of applied genetic knowledge on large numbers of people.

Expanding the Impact of Genetic Knowledge: Mass Genetic Screening

That the emphasis has so far fallen most heavily on ridding mankind of genetic disease should not obscure the fact that the vision of genetic improvement has a lively life just below the surface, in the stirrings of a new eugenics movement.
—Daniel Callahan, *The Tyranny of Survival* (1975)

Genetics has moved from the small-scale laboratory exercises of a few pioneering scientists to massive experiments conducted on whole populations. In the last ten years, close to 20 million persons have been subjected to some form of genetic test. This transition has meant that the individual attention and care usually afforded to persons in the doctor-patient relationship have had to give way. In their place one finds the kind of anonymity and neglect of human rights that may accompany any mass movement. The change in scale has been accompanied by a subtle change in perspective as well. The intensely personal concerns of the individual about his or her genetic deficits has telescoped outward, giving us a new view of society's problems and imperfections.

The vanguard of this movement has been called *mass genetic screening*. Genetic screening programs look for chemicals in the blood or urine that signal the presence of underlying genes whose products cause or can predispose to disease. Other screens search for genetic markers that indicate the likelihood of future physical difficulties, such as emphysema (associated with defective gene products associated with lung damage); or immunological deficiency (associated with genes that code for an enzyme known as adenosine deaminase). Still other screening programs may search for chromosomal abnormalities in unborn fetuses, with the aim of identifying the kind of future difficulties that convince prospective parents of the wisdom of abortion. As in biochemical screening, the degree of certitude in these programs ranges from absolute assurance of massive disorder (for example, where whole chromosomes or chromosomal segments are missing) to vague storm warnings of future difficulty (for example, where an extra long Y chromosome is found).

Some biomedical and health researchers have hailed the expansionist spirit in modern genetics as a welcome boost for preventive therapeutic approaches—that is, for procedures to detect, identify and treat disease processes before the symptoms appear. Other observers have questioned the wisdom of such programs, both for individuals and for the target populations involved in mass screenings. Friends and critics alike, however, must consider the difficult therapeutic questions raised by the advent of massive genetic screening networks.

Should screening programs be guided by what Talcott Parsons has defined as the principle of "social efficiency," according to which the genetic deficiencies which are most "costly" in terms of social or work efficiency are given priority? Should screening be designed to identify those persons most *in need* of medical care, without regard to the prognosis for useful functioning? Or should we try to dovetail genetic screening with other social and medical objectives?

The ultimate dilemma of genetic screening is that it gives us a technological ability to implement seeming solutions to human problems before we have a full understanding of their social implications. We now have the technological ability to detect literally hundreds of compounds in a single exhalation of breath, and to analyze body chemicals whose presence was only inferred just ten years ago. But what we should *do* with that information is much less clear.

A history of the health-screening movement generally shows that we often pressed forward with testing programs before we had resolved the question of the adequacy of our therapeutic options, or the impact that such testing might have on the choices available to consumers for controlling their own health care. TB screening, breast-cancer testing, and cervical-cancer screens all at one time or another were out of synchrony with the long-term health benefits that they promised, and with our ability to deliver them—to say nothing of the possibility that the testing itself might do more harm, or cost more, than the disease being sought.

From its primordial origins in tests for elevated levels of blood phenylalanine (PKU testing), mass genetic screening has followed a similarly checkered history. We now know that such screening carries hidden as well as overt costs and benefits. And as we have seen, genetic information itself is often heavily charged with psychological and social implications for the consumer.

With rare exception, these implications have not been adequately considered in the inception or design of genetic screening programs. Instead, most such programs are rationalized on the ground that they are a logical part of the movement to "prevent" disease or to broaden the purview of science itself. Thus, the net of genetic analysis is thrown further and further afield from appropriate scientific and therapeutic objectives. What was once research into the nature and extent of human genetic variability may now crop up as a rider to a bill funding speculative inquiry into social or environmental problems.

In recent years, for instance, tests have been applied to newborns to identify those with genetic characteristics (e.g., XYY sex-chromosome makeup) thought to be indicative of negative social behavior (criminality). Similarly, adults have been screened prior to job placement to determine their "susceptibility" (as suggested by, for example, their *Pi* alleles) to diseases such as emphysema known to be caused by occupational hazards. To the extent that such associations are presently untested and unproved, a "genetic predisposition" test is no different in kind from the assessment of racial type as an indicator of resistance to hazardous solvents or other chemicals (blacks, of course, were presumed to be more resistant than whites, and so on). Other proposed screening would have adults tested for the purported but unproved genetic basis for susceptibility to lung cancer.

Some of these efforts are based on fallacious assumptions about genetics and about pathology; others exemplify the kind of confusion of purpose that inevitably follows a confusion of values. Clearly, the design of genetic screening programs ought to be tailored to the program objectives, to the techniques employed and, above all, to the needs and well-being of the program participants. All too often, however, genetic screening programs have not met these design goals. Historically, they have blindly copied screening programs for the nongenetic diseases and have overlooked both the special demands of genetic research and the special requirements of participants in such research programs.

Currently, the intensity of these problems differs with the type of screening program undertaken. Most genetic screening programs can be seen as falling into one of three categories: presymptomatic screening, parental screening and research screening.

In *presymptomatic screening,* a potential victim of a genetic disorder is sought out before any symptom of the disorder becomes evident. The objective here is to identify such individuals in time to initiate preventive treatment before any irreversible damage has been done. Phenylketonuria (PKU) testing of newborn infants along with its newer counterpart, congenital hypothyroid screening, is the archetype of presymptomatic genetic screening; but childhood screening for visual and hearing defects, and screening for diabetes, glaucoma or hypertension can all uncover previously unsuspected genetic contributions which technically make them presymptomatic genetic screens. There are, at worst, major conceptual and operational differences between genetic and nongenetic screening techniques, based on the fact that genetic disease raises the psychological specter of an irremediable process.

In *parental screening* the objective is to help prospective parents determine whether a future child is at risk for a particular genetic disease. Tay-Sachs testing, sickle-cell-carrier testing, the testing of pregnant mothers for alpha fetoprotein (for neural-tube formations), or for parents who may be translocation carriers for fetal chromosome abnormalities, are all examples of parental screening. Prenatal screening is not unique to the genetic disorders. Premarital detection of blood-group incompatibilities, venereal diseases or prenatal detection of microbial infections are analogous situations. Because it involves issues of heredity and parenting, however, parental screening is also likely to involve volatile psychological issues.

Carrier-screening is a major form of parental screening which may be extended to younger age groups. Recent studies of groups subjected to this kind of genetic screening seem to indicate that conferring such genetic information on individuals may degrade their self-esteem and set off accompanying psychological difficulties. For example, nearly half the participants in a genetic screening program who learned that they carried the gene for Tay-Sachs disease (a classic recessive condition that causes early and profound neurological degeneration) were emotionally upset by the information, even though they were *not* married to another carrier, and hence not at risk for having affected children. In this situation, some screeners concluded that the psychological stress to the carrier was so great that it was not worth the prospect of finding two who were spouses and hence at risk for transmitting the gene in a disease-conferring form to their offspring.

A stress reaction disassociated from the possibility that one could actually transmit the disease was almost totally unanticipated by the earliest screeners for genetic disease. Nor did early screeners anticipate that noncarriers would stigmatize those found to carry a single dose of the gene being sought. A study of the aftereffects of a sickle-cell–anemia screening program revealed that the genetic information itself, rather than other possible causes, seemed to be the source of distress in carriers. Participants who "escaped" identification as carriers showed much lower levels of anxiety, depression and hostility that were characteristic of the newly discovered carriers. Similar results were reported in a Canadian study in which schoolchildren found to be carriers of Tay-Sachs disease suffered from anxiety and depression, and were socially ostracized by their noncarrying schoolmates.

Perhaps more importantly, the addition of genetic counseling to neutralize or buffer this stress proved largely ineffective. And parents who were told that their offspring had the inconsequential single dose of the gene (i.e., one virtually incapable of producing pathological effects) nonetheless were upset at learning so indirectly of *their* flawed genetic makeup, since at least one parent had to be a carrier to have a carrier offspring.

Studies of a population of blacks who went through screening and counseling showed that a significant proportion, after the counseling session, still believed, erroneously, that carrying a single dose of a sickle-cell gene (sickle-cell trait) itself was some kind of disease state; that vigorous play must be restricted for children with the sickle-cell trait; that special dietary supplements are needed for treating the trait; and that having it will result in some adverse physical manifestation. Taken alone or collectively, these reactions betray the strong forces of denial that unacceptable genetic labels can engender.

Pilot studies conducted at Howard University with a sociological measure of self-esteem repeated these general findings: black college students who had gone through screening expected a clean bill of health; when they emerged with the diagnosis of being a carrier, some felt less powerful and controlling than when they went in.

Thus, both Tay-Sachs and the sickle-cell studies seem to indicate a previously unrecognized psychological cost to screening, independent of any practical effects of knowledge of the genetic contribution to their biological makeup.

In *research screening,* the objective is to collect data for eventual determination of the appropriateness of presymptomatic or parental screening. Here, the participants or their parents may be given only unproved information of doubtful interpretation. Testing for hyperlipidemia in childhood, for XYY males at birth, for Huntington disease in adolescents, or prenatal detection of hemoglobin variants of uncertain prognosis, are all examples of research screening. Initially, some of these programs were designed simply to explore a new frontier, others to provide early study of the feasibility of "therapeutic" interventions, as in the prenatal diagnosis and abortion of a child with sickle-cell anemia.

As an example of the flaws that come from incomplete planning, consider the aftermath of the first reports of an extraordinary prevalence of an extra-Y chromosome among the male inmates of a mental penal institution in 1964 and 1965. Within a few months we were told that researchers had found a gene that determines criminality. Only seven years later, in Massachusetts, the Law Enforcement Assistance Administration sponsored a study that was intended to develop a way to quickly detect prospective "extra-Y" males from their palm prints. Three years later, more than fifty studies had been conducted without anyone developing an adequate conceptual model of how an extra-Y chromosome, as distinct from an extra-X, might cause criminality, even though both chromosomal constitutions were equally prevalent in mental and penal institutions. Later studies showed that the sampling of the original studies was flawed, and that the association of purely physical changes or the mental retardation that accompanies the extra chromosome may be the primary cause of the increased risks of conviction and incarceration of affected individuals.

The reported higher frequency of at least two major sex chromosomal abnormalities (XXY and XYY) in institutions for the criminally insane led to an increased emphasis on the need to obtain base-line data on newborn incidence against which to measure the significance of the apparent concentration in institutions of adults having these chromosome makeups.

In two articles describing males with the XYY syndrome, the authors emphasized the difficulty in assessing the possible contributing effects of the psychological setting in which these males lived. Danish workers commented on the problem of stigmatization and biasing of test results generated by the presence of an observer and the overt identification of "affected" individuals. This and other objections to the paucity of controlled studies led to the call for prospective studies to identify newborns with chromosome variants and follow their psychomotor and linguistic development through early life.

The problems that mass newborn cytogenetic screening might generate are highlighted in an article in the August 1974 *Pediatrics.** Subsequent to the discovery of eleven chromosomally abnormal individuals in a survey of 4,400 consecutive newborn infants, a group of genetic researchers followed the psychomotor development of the normal and genetically abnormal children. Noting that no distinguishing clinical abnormalities could be discerned in the variant population, the authors concluded that "one cannot predict developmental potential from a knowledge of the genetic constitution." These impressions are at variance with previous reports and thus serve to emphasize the need for more definitive work. Such discrepancies highlight the question of whether it is *possible* to design, for research in this area, criteria that adequately take into consideration such problems as the subtleties of the parental role in child rearing or the problem of stigmatization compounded by a self-fulfilling-prophecy effect of the widespread association of an extra Y to criminal tendencies. Irrespective of these difficulties, a very large number of prospective surveys of cytogenetically distinguishable

*M. F. Leonard, et al., "Early Development of Children with Abnormalities of the Sex Chromosomes: A Prospective Study," *Pediatrics,* Vol. 54 (1974), 208–212.

variants among newborns were organized in the late 1960s and early 1970s, including some in which behavior-modifying intervention was contemplated for dealing pre-emptively with anticipated behavioral deviance. By 1975 most studies had been halted.

In surveying the contemporary screening landscape, one finds considerable attention paid to confounding variables like these in the designs employed in diverse experimental and therapeutic situations. The common ancestor of the contemporary forms seems to have been the presymptomatic format employed in the PKU screenings of the early 1960s. Given the deluge of genetic screening programs that followed PKU, an understanding of the history of this particular research and treatment effort helps to illustrate how flaws occurred.

PKU Screening Programs

In 1963 a cheap and presumably accurate test for detecting excess amounts of phenylalanine in the blood was introduced. Although only a rough, qualitative indicator of the disease potential of the individual, the test (known as the Guthrie test) was inexpensive and widely accessible. Since the dietary treatment for the condition, though unproven, was also readily available, PKU emerged as the first genetically determined condition for which there was both a ready means of identification and treatment.

Also in 1963, Massachusetts legislators introduced the first state law providing for compulsory testing of all newborns for PKU. Barely a year and a half later, some thirty states had passed similar laws. Some states had mandated and funded treatment programs as well; but most states funded testing only, leaving many individuals without the financial resources to pay for treatment. By

1974 the number of states mandating testing had reached forty-three.

Despite this very rapid governmental endorsement of PKU testing, many authorities harbored doubts about the testing program. One such expert, Dr. Harold Nitowsky, of the Downstate Medical Center in New York, expressed the substance of these reservations concisely:

> I believe that we shall be forced to the conclusion that our knowledge of the natural history and variability of PKU is incomplete, that the effectiveness of treatment of the disease has not been accurately measured, that we have inadequate information about the optimal age for institution of therapy, or the levels of serum phenylalanine at which treatment should be undertaken, or the age at which treatment may be stopped.

Dr. Nitowsky concluded:

> Despite these unanswered questions, and the obvious lack of adequate validation of prescriptive screening, I do not believe we should turn backwards. Our intuition and empirical judgments would deter us from altering current practice.

Like Nitowsky, most researchers have chosen to move ahead with PKU testing. As a result, a whole generation of children has been destined by law to participate in PKU experimentation; and part of this generation has served as an experimental population upon which specific combinations of test reagents, dietary treatments and assay techniques have been worked out. Even infants from non-European backgrounds, whose chances of developing the disease were virtually nil, have been destined to participate. This inequity has led the District of Columbia legislators to suspend PKU screenings in 1971, after three years of testing and an outlay of $100,000 had failed to identify a single affected newborn. Similarly, between January 1976 and July 1977, only one case of PKU was successfully diagnosed out of 136,370 infants screened in Maine, Massachusetts and Rhode Island. While still not at the expected level of case finding, the experience of other states has been more productive. Between 1966 and 1974, for instance, California health workers screened almost 3.5 million newborns and found 147 who had persistently high phenylalanine levels in their blood.*

These seemingly favorable results obscure the reality of PKU screening. Even under the most optimal conditions of screening, in which the sensitivity of the test for detecting raised phenylalanine levels is 100 percent, and its specificity for distinguishing phenylalanine is 99.95 percent, the actual predictive value for finding a case of a child who would otherwise undergo mental retardation from frank phenylketonuria has been calculated by medical researchers to be no more than 17 percent of the total. In fact, according to genetic-screening expert Dr. Neil Holtzman of Maryland, a positive diagnosis is confirmed only in 5.1 percent of the infants whose PKU values were originally scored as "high"; more than 10 percent of children who actually have PKU (circa 1970) were not being detected at all; and the number of false positive cases have been increasing.

Part of the blemished picture of PKU screening can be traced to the overly sanguine and simplistic views of the first screeners. It was widely believed in the early days of screening that the treatment itself was harmless and uniformly effective—and, more egregiously, that through treating PKU, there could actually be a society-wide savings in terms of a net reduction in mental retardation. All three of these beliefs have since been shattered, the last

* In contrast, congenital hypothyroid screening yields about 1 treatable case for every 4,000 newborns tested, or about 6 times more cases per test than PKU.

because the children of PKU mothers are almost always born retarded—and 2 out of every 3 of their well children perpetuate the gene in its carrier state. While researchers are constantly reviewing and updating their treatment techniques, for instance to assure that growth retardation does not follow dietary treatment, 100 percent safety and efficacy (i.e., restoration of full intellectual potential) is still not within reach.

Many of the difficulties of PKU testing could have been anticipated if screening programs had been more carefully thought out. At the time when compulsory legislation was under study in many states, clinicians were beginning to recognize that PKU was, in fact, several different disease processes. As early as 1966, novel variants of the disease were known to occur. Since then a whole spectrum of genetic disorders has been found, including symptoms characteristic of the traditional PKU child. Indeed, in the last few years, some patients have appeared with the genetic characteristics of PKU, but with symptoms so mitigated by either environmental or secondary genetic effects that the so-called "affected individuals" are in no sense clinically sick.

Unfortunately, the negative effects of such testing are not limited to the misappropriation of research and treatment resources. The PKU diet, if applied to a normal child, can itself induce major metabolic disturbances affecting growth and development. A case in point is that of a Maryland child who was diagnosed as having PKU at birth and subsequently was placed on the prescribed diet. Under ordinary circumstances, the restriction of circulating phenylalanine levels accomplished by the strict diet would have been more than sufficient to control the child's genetic problem. But by age seven, this child had developed mental retardation—a condition that the diet had been designed to circumvent. Careful retesting showed that the child did not have classic PKU at all, but rather a related metabolic defect (a deficiency in an enzyme known as dihydropteridine reductase), which produced similar test results.*

The history of PKU screening programs suggests some important lessons that may apply to genetic screening generally. In the first place, the rapid proliferation of compulsory PKU-testing programs is suggestive of the political influence wielded by proponents of a geneticist view in our society.

Secondly, these proponents are naturally concerned with the well-being of their own research programs. One indication of this concern is the public-relations effort that has accompanied the institution of compulsory testing for other metabolic genetic diseases. Unlike PKU, these programs have proved to be more akin to biomedical research than to mass treatment.

For political and financial reasons—as well as misguided public-health intentions—it has been expedient to launch widespread genetic screening programs under the banner of therapeutic interventions when in reality many are at best tentative probes to measure the extent and variability of gene-associated diseases. Were these programs billed as such, their limitations as public-health programs would be more acceptable, and the possible physical and psychosocial costs to the participants would be more readily offset by appropriate counseling. And when a therapeutic intervention was first being tried, it would be better all around to identify its research nature, so that parents might be apprised of benefits and risks in advance of treatment and thereby given the opportunity to object.

Thirdly, the long-term therapeutic effects of the PKU programs have been, at best, dubious. On balance, PKU

* A separate test for this abnormality is now being instituted in most states.

screening has probably helped fewer persons at risk for mental retardation than would have been helped had a similar effort been made in a preventive health screening program that measured hypothyroidism. More to the point, PKU screening could probably have been launched with less harm to participants and a greater benefit risk ratio all around, had it been confined to those families who had a history of PKU. Indeed, some medical researchers have calculated that fully half the cases of PKU detected with blanket tests of the entire newborn population (culling 300 cases for every 10 million screened) could have been found by limiting screening to the at-risk population. With the advent of reliable adult-carrier tests for PKU, this classical approach to disease intervention is a near reality, since only those couples where both partners were PKU carriers would have to have their newborns tested for elevated phenylalanine.

Finally, the PKU experience suggests that the planning of screening and interventions for even the most "classic" gene-caused metabolic disease might profitably be subjected to renewed scrutiny.

Other Screening Techniques

Despite the PKU experience, genetic screening programs have grown in number and complexity through the 1960s and 1970s. Today there are screening techniques available that earlier researchers might only have dreamed about. Among these are the following:

Multiphasic screening permits the use of one or more samples of body fluids to ascertain a profile of critical metabolites or other substances that might indicate a genetically based prediiection for disability. The Massachusetts metabolic screening program is a representative example.

In Massachusetts some fifteen genetic abnormalities are sought from blood or urine specimens taken from the newborn. The Massachusetts program uses three separate tests: (1) an umbilical-cord blood sample taken at birth to screen for such potentially lethal diseases as galactosemia (where the earliest possible detection is vital); (2) a peripheral blood sample taken soon after dietary protein has been introduced (usually two to four days after birth); and (3) urine samples taken when the newborn infant is three to four weeks old. The last approach is favored by some researchers because separation and identification of compounds in urine by paper chromatography can identify abnormalities of membrane transport, which would be missed if only the blood were examined. In the mid 1970s, for instance, approximately 260,000 urine tests were performed in Massachusetts each year, an average of about three tests per patient. The "extra" testing was required to establish the significance of any abnormal result by eliminating false positive results due, for example, to bacterial contamination. A number of consecutive positive tests were required for confirmation of a metabolic disorder requiring further investigation.

In its current form, the Massachusetts program screens approximately ninety thousand infants annually and detects approximately thirty-five infants per year who do have some metabolic disorder. Of these, about 60 percent subsequently manifest clinically significant disease. The Massachusetts program costs about $200,000 per year—or about $1.75 to $2 per infant screened. About 80 percent of the program's annual budget comes from a federal grant (Children's Bureau, Health Services and Mental Health Administration), the remaining 20 percent from the Massachusetts Department of Public Health and individual payments.

Recent technical developments in biomedical engineering are beginning to make mass multiphasic testing appear technically feasible and increase the likelihood of widespread application. Automated tests are available to detect at least eight significant metabolic disorders: PKU, maple-syrup urine disease, homocystinuria, histidinemia, valinemia, galactosemia, argininosuccinic aciduria, as well as congenital hypothyroidism. While most current tests rely on bacterial growth-inhibition assays of the type used in the Guthrie test, other developments that may facilitate automation are on the horizon.

Prenatal screening allows the detection of genetic abnormalities at a time (fourteen to twenty weeks) when abortion is feasible. This type of screening thus presents the greatest challenge to public-policy makers, since many citizen groups are concerned over the expansion of indications for abortion and the subtle coercion which wholesale public participation might bring. Nevertheless, the medical literature has increasingly dealt with the utility and versatility of prenatal diagnosis for the detection of chromosomal anomalies, a variety of single-gene defects and neural-tube defects. Most of these diagnostic procedures entail amniocentesis—the withdrawal of fluid from the sac surrounding the fetus by means of a needle insertion. Two United States collaborative studies on amniocentesis, involving over four thousand cases, has shown this technique to be reasonably safe. Following any successful amniocentesis procedure, unfavorable information about the fetus would afford the parents and attending physicians specific data that can be used as a basis for an abortion decision.

It is now common at major medical centers for pregnant women past the age of thirty-five to be offered amniocentesis because of the increased risk of bearing a child with a chromosome abnormality (specifically Down syndrome) in this age group. A formal proposal for such chromosome screening, beginning with women over thirty-five and gradually becoming a routine part of prenatal care for all pregnant women, has already appeared and has been strongly endorsed by the California legislature and by the Acting Director of the National Institute of Child Health and Development.

Maternal-blood testing is another technique available to physicians and patients interested in the genetic characteristics of the unborn child. It has been used to detect women at risk for Rh antigen sensitization and to thereby anticipate fetal hemolytic disorders. Maternal involvement has been supplemented by the development of a blood test that can suggest neural-tube defects in the fetus. Pioneered by Brock and his colleagues in the early seventies, this "alpha fetoprotein" test has been developed to the point where it allows a rapid initial screen of high-risk populations. Following a positive serum test, an amniocentesis must be done to confirm abnormal alpha fetoprotein levels. But, as measured by the occurrence of a high proportion of unanticipated spontaneous miscarriages and misdiagnoses from subsequent amniocenteses, the human "costs" of such programs appear to be high. (This view is hotly contested by advocates who assert that miscarriages would have occurred anyway—an untestable proposition for any individual case.)

The Sickle-Cell Programs

Perhaps the best-known examples of genetic screening programs are those that attempt to identify carriers of sickle-cell trait and persons with sickle-cell anemia.

Persons who have sickle-cell trait have a single gene that differs from the normal hemoglobin gene in but one of 438 possible base sequences. Persons with the trait express this gene along with the normal one and hence have *two* kinds of hemoglobin in their red blood cells: hemoglobin A, which is comprised of two "alpha" and two "beta" polypeptide chains; and hemoglobin S, in which one of the alpha chains has the sickle cell amino acid substitution. Because the S type is present in only about 20 to 40 percent of the total hemoglobin, these persons function just about normally in every respect: their life expectancy is the same as persons with all normal hemoglobin (and even longer where tertian malaria is present); their physical and mental abilities are indistinguishable; and so on. It is only in conditions at the extremes of human adaptation, such as unassisted high-altitude mountain climbing and total anesthesia, that any signs of difference show up. Thus the hemoglobin of persons with sickle-cell trait works for all practical purposes as well as does that in persons with the more common gene combination of pure hemoglobin A.

But when *both* genes for hemoglobin S are present or, less commonly, when another variant is present, the person is in trouble. Sickle-cell anemia is a disease. It draws its name from the peculiar shape taken on by the red blood cells when their internal structure collapses. Cells "sickle" when the normally oxygenated hemoglobin loses its oxygen and structural supports in the protein are disturbed, thus permitting the molecules to crystallize. It is believed that this unique distortion of the hemoglobin molecule helps to make hosts with one or both S genes particularly unpalatable to malarial parasites. Other hypotheses abound, but the net results have been there for all to see once it became possible to visualize the different hemoglobins: Those persons who carried hemoglobin S (and certain other forms like hemoglobin C) survived while those with normal hemoglobin in their blood cells succumbed wherever the malarial mosquito thrived around human habitations.

Because malaria used to be endemic in the Mediterranean area and parts of Africa and Asia, malaria's depredations there increased the number of persons who had the protective genetic makeup characteristic of sickle-cell trait carriers. The gene for *thalassemia,* a related hemoglobin abnormality, is another which increases the survival capacity of carriers. As a result of this classic form of natural selection, these genes are found more frequently among nonwhite minorities in America who originated in the Mediterranean basin. During the late 1960s, massive infusions of state and federal money helped to establish screening programs all over the country. For many black legislators, this effort served as a kind of symbolic redress for the gross inadequacies that have otherwise characterized health care for many nonwhite Americans.

Between 1970 and 1972 at least thirty states initiated or approved legislation for sickle-cell-anemia or sickle-cell-trait screening. At least eleven made such screening mandatory. In Washington, D.C., sickle-cell anemia was legally declared a communicable disease, and testing for schoolchildren was made compulsory. In Massachusetts, sickle-cell trait was legally declared a disease, even though it is the virtually innocuous carrier state of the actual recessively inherited disease. And in New York and a number of other states, blacks were asked to undergo sickle-cell testing when applying for marriage licenses.

It was in the attempts to abate this early rush of legislation that the first references to the political implication of genetics came about. Six states have since reversed or repealed their mandatory laws (New York, Illinois, Massachusetts, Maryland, Georgia and Virginia) but not before the Armed Forces began mandatory testing of recruits and several airlines discharged persons with sickle-cell trait from flight duty. An unknown number of blacks with sickle-cell trait were refused health insurance or employment or were charged increased rates for coverage.

The story behind the movement that began to reverse the tide of genetic politicization is short and worth retelling if for no other reason than to show how future abuses may be buffered and eradicated.

5

Legislative Remedies to Abuses of Genetic Knowledge

The future battles for the modification of man will undoubtedly be fought in the judicial arena . . . If the individual's right to be left alone is to be protected, the searchlight of public scrutiny must be focused upon this long-ignored power to modify man. The traditional Anglo-American tools of judicial process and review provide some of the most effective means for scrutiny known in any social system.
—N. N. KITTRIE, *The Right to Be Different* (1971)

In September of 1971 I took a job with the Institute of Society, Ethics and the Life Sciences, in Hastings, New York. As director of the Genetics Research Group, I planned to dispose of the problems relating to the abuse of genetic knowledge in short order and move on to other business. But the problems were a bit larger than I had bargained for, and in 1978 I still found myself struggling to develop suitable legislative remedies against the prospect of abuse of genetic knowledge.

By December of 1971 I had begun to realize the

dimensions of the abuse of genetic knowledge—real and potential. Genetic "discoveries" were proliferating over a broad front, spawning enormously complex ethical and political problems along with them. It was then that Massachusetts passed its legislation treating sickle-cell anemia (the prototype of later "genetic diseases") as if it were a communicable disease warranting emergency public-health measures to combat it. In that same year New York was considering legislation that would require non-whites seeking a marriage license to receive testing for sickle-cell trait. In each instance, members of the black community unwittingly pressed the cause of legislative action. These movements presaged a strength and political weight that gained a strong foothold in legislative bodies throughout the nation for identifying carriers.

In the midst of this sickle-cell "epidemic" in December 1971, a small group of geneticists and ethicists met at New York's Biltmore Hotel. I was presenting a preliminary set of guidelines to the group—which included James Gustafson, of the University of Chicago; Richard Roblin, of Harvard; John Fletcher; Sumner Twiss, of Brown; and Tabitha Powledge, of Hastings—when Dr. Robert Murray, Jr. strode in, visibly upset. Murray had just come in from Washington, where the city council had decreed that sickle-cell anemia was a full-scale communicable disease. In spite of the fact that genetic disease differed radically from such communicable diseases as smallpox, Washington, D.C., had acted as if sickle-cell anemia posed a life-threatening risk to the community. Murray felt strongly that sickle-cell anemia did not warrant the kind of emergency (and highly intrusive) screening and treatment techniques traditionally reserved for communicable disease epidemics. In his view, genetic disease was being misused as an excuse to suspend the right of free choice, and to breach the traditional privacy of the doctor-patient relationship by making information produced by genetic screenings widely available. Murray's anger and concern swept through the group, and we began to work in earnest on a set of guidelines to regulate genetic screening programs.

Two months later, the group issued a draft statement comprising eleven principles:

GOALS. Screening programs should have clearly designed goals. They should be capable of presenting evidence (through pilot projects and other studies) that the program goals are achievable.

COMMUNITY PARTICIPATION. Program planners should attempt to involve representatives of affected groups in the program's design, administration and evaluation.

ACCESS. Information about screening and screening facilities should be open and available to all. Education can be a program's most important therapy.

TECHNIQUE. Some past screening efforts have produced misleading information because of inaccurate or imprecise testing procedures. Such procedures not only should be free of technical difficulties, but also should ideally be designed to provide maximal information minimally subject to misinterpretation.

FREE CHOICE. The group strongly urged that no screening program attempt to impose constraints on childbearing by individuals of any specific genetic constitution. "It is simply unjustifiable to promulgate standards for normalcy based on genetic constitution," the group concluded.

INFORMED CONSENT. As in all other forms of research employing human participants, genetic research demands the informed consent of its subjects. Increasingly, genetic programs are focusing on minors; and in this case, the group urged experimenters to avoid designing programs for preschool or preadolescent participants because of the high risk of stigmatization, and the low medical value of such studies. Instead, the group urged neonatal testing to maximize the therapeutic value of counseling and other techniques. Program officials are accountable for seeing to it that the informed-consent provision is met.

HUMAN RIGHTS. Because most techniques for the identification of the genetic base of a particular biological process are novel and experimental, genetic screening is governed by rules and codes regulating "human experimentation." The Department of Health, Education and Welfare, among others, has published guidelines for such studies that minimize harmful effects on participants.

FREE INFORMATION. Participants in genetic screening programs almost always have more than a passing interest in the outcomes of such research. Accordingly, programs have an obligation to disclose all unambiguous diagnostic results to the person, his legal representative, or an authorized physician. General results ought to be made available to the community.

COUNSELING. Programs must be prepared to deal with the human consequences of the information they produce. This means provision of well-trained genetic counselors to assist those identified as heterozygous (or, more rarely, homozygous) for a particular trait. Counselors should inform, then listen; they should not try to decide for their clients.

TREATMENT. Anyone identified as "having" a particular genetic constitution is very likely to want to know what can be done about it, should the constitution be seen as deleterious. Information about available therapy—its nature, cost and effectiveness—should be made available before screening proper. A knowledge of treatment possibilities may be seen as a condition of "informed consent." Similarly, consent to participate in *treatment* is a separate question from consent to participate in *screening*.

Privacy. Most states do not have laws recognizing the confidentiality of public-health information. Consequently, the group favored a policy of informing only the person to be screened (or, with permission, a designated physician or medical facility) of the study outcomes. Moreover, the group advocated special record-keeping provisions for securing the information, including coding of records, and prohibition of storage of uncoded information in data banks where telephone computer access is possible.

Following adoption of the guidelines by our subgroup, the next step was to put some punch into them by obtaining wider endorsement. Eventually, all twenty-odd members of the Genetic Research Group—scattered geographically between both coasts and ideologically all over the map—signed the guidelines. After their publication in the *New England Journal of Medicine*,* these guidelines became a focal issue in debate between scientists and policy makers, and a touchstone for groups and health departments nationwide. California and Maryland have since introduced regulations based on these principles and have put legislation into effect to give these provisions the force of law.

Nevertheless, the provision requiring informed consent attracted the fire of critics. They pointed out that in some cases where therapy is available, neonatal or prenatal

* M. Lappé et. al., "Ethical and Social Issues in Screening for Genetic Disease," *New England Journal of Medicine*, Vol. 286 (1972), 1129–32.

testing is the only way to maximize the child's chances for successful treatment. For example, denying a child with a predisposition toward PKU a chance to be tested because his parents either refuse consent, or aren't available to supply it, might result in mental retardation. In this way, a child (obviously incapable of giving consent or refusing it) might be subject to a life of deprivation because of the vagaries of circumstance or parental will. Since publication of the guidelines, I've tended to agree in part with these critics, and favored recognition of the legitimacy of permitting state-mandated testing, where treatment can prevent or ameliorate the development of otherwise debilitating malfunctions—as long as the parent is informed of the benefits (and risks) of testing and is given a reasonable opportunity to object. (In California we have successfully changed our policy to honor parental refusal.)

The *New England Journal of Medicine* article also sparked a legislative response to the potentially harmful effects of mass screenings. Soon after publication, for example, the National Sickle Cell Anemia Act of 1972 was spawned and Maryland drafted first a law governing sickle-cell, then a comprehensive "hereditary disorders act." The Maryland Law (Chapter 693 of Maryland Statutes for 1973) established a sixteen-member commission on hereditary disorders charged with regulating genetic screening programs. The commission was unique in at least one regard: it consisted of a majority of lay members who, in addition to doctors and members of the state health department, could exercise extraordinary powers of control over the conditions under which citizens were subjected to genetic testing. The commission may control the manner in which genetic information is recorded and released, investigate charges of discrimination resulting from the abuse of such information, and mitigate future abuses by exercising strict controls over the confidentiality of records.

I had a lesson in the importance of such controls from the president of the American Heart Association in New York, who was interviewed early in the process of drafting guidelines. He argued that insurers had an absolute right to minimize the financial burden they assumed in underwriting high-risk clients. I pointed out that this didn't seem relevant to the question of making sickle-cell test results available to insurers, since actuarial tables showed no increased death rates for sickle-cell carriers. He replied that black males were certainly at higher risk than other insurable groups, and that the sickle-cell information would help identify blacks to potential underwriters(!).

The principles originally proposed by the Genetics Research Group were formally incorporated into model legislation at the annual Council of State Governments conference in Seattle in 1973. California Senate Bill 747 (Chapter 1037 of California Statutes of 1977) incorporates many of these provisions. Introduced by Nate Holden (D-Los Angeles), this bill is intended to minimize the psychosocial injury to persons with hereditary disorders and their parents. The bill calls for consultation with community groups affected by hereditary disorders prior to the institution of any screening program. It also requires public-health education as to the nature of the genetic disorder, protection of the confidentiality of screening data, full disclosure of the information and the uses to which data will be put to screening participants, and the provision of counseling services to participants found to be at risk.

Like the Maryland bill, the California bill breaks new ground in prohibiting the testing of a minor over the objections of his or her legal guardians. It assures that no test program may require restriction of childbearing and guarantees that eligibility for social services shall not hinge on participation in screening or treatment programs. Perhaps the most significant departure of the California bill, however, is its exemption of neonatal programs from informed-consent provisions. Mandatory screening for PKU and galactosemia is thus allowed under the California bill, with the provision that persons found to be "affected with genetic disease be informed of the nature and, where possible, the cost of available therapies or maintenance programs, and be informed of the possible benefits and risks associated with such therapies and programs."

Newborn programs were in place and operating at the time this legislation was being written in hospitals throughout California. Personnel in the State Department of Health's Genetic Disease unit believed that placing informed-consent requirements on such programs could create enormous administrative burdens, as well as possibly unsupportable delays in service delivery. On the other hand, retaining the consent provision would allow parents to learn about the purposes of the program, and to participate in the decisions that would potentially affect their children's lives. As the Maryland experience indicated, consent could increase parental cooperation in the drawn-out process of testing and retesting necessary to confirm the initial diagnosis.

Clearly then, a compromise was called for, and a compromise eventually emerged, giving parents the right to know about (and to prohibit) tests performed on their children. Prior consent, however, was not demanded. The relevant section read: "No test or tests shall be performed on any minor over the objection of the minor's parents or guardian, nor may any tests be performed unless such parent or guardian is fully informed of the purposes of testing for hereditary disorders, and is given reasonable opportunity to object to such testing." This provision honors parents' rights, while allowing the state to discharge its obligation to provide services to all newborns.

At best, the law is a slow and imperfect means of

problem-solving. In Maryland and California, however, the law now affords some protection to those in need of it because of "hereditary disorders." These laws accomplish two major goals: First, by providing education and counseling services, they help to demystify genetic knowledge and detoxify data from genetic screening. Second, such laws afford individuals protection from social stigmatization and exploitation. They prevent insurers, for example, or governmental bodies or researchers from using genetic data to characterize at-risk individuals for other than therapeutic purposes.

State legislatures can and should use their authority to counterbalance the enormously powerful financial, racial and even scientific pressures to employ genetic arguments in a self-interested way. N. N. Kittrie has suggested that such arguments take an a priori approach to social problems. Society deemphasizes the importance of the individual's responsibility for actions in favor of treating such problems at the universal source—the gene. As Kittrie has observed:

> If legislatures remain cautious, the authority for social experiments in human modification will rest with the administrative level, even though it is highly undesirable for drastic state power over procreation or other human modification to be invoked by experts and administrators without prior legislative deliberation and enactment.

As knowledge about the genetic underpinnings of disease processes continues to expand, the issue of who "owns" genetic data will certainly become an explosive political issue. In California and Maryland, state legislators have already acted to insure that individuals and groups are protected against the misuse of private information by genetic enthusiasts. The problem to be confronted in other states is not whether genetic knowledge ought to be controlled—for someone, clearly, will control it—but who shall control it and to what ends.

Thus, the early rush of poorly constructed legislation has largely abated, and the future prospect of heavy-handed laws has been greatly reduced by these legislative remedies. The capstone to this process was the National Sickle-Cell Anemia Act of 1972 and the Genetic Diseases Act of 1976. Both mandated that screening be done only under the most scrupulous conditions, and that it always be voluntary and unlinked to eligibility for other federal services. This last provision strikes to the heart of the prospect that genetic screening might be used to foster eugenic programs, as was so strongly possible when sickle-cell screening was placed under federal aegis at the centers for family planning.

But a residue of doubt remains as to the wisdom and motivations of those who will continue to participate—and even run—mass genetic screening programs. If we are to believe the early results of sociological and scientific surveys, we still have strong reason to question the value orientations and knowledge base of genetic practitioners. Some medical practitioners have maintained conflicting and often ill-informed opinions as to the medical significance of sickle-cell anemia or trait. Wide differences in interpretations of the severity of both conditions were revealed in a study conducted by James Sorenson while he was a member of the Genetics Group that I directed at the Hastings Institute of Society, Ethics and the Life Sciences. Sorenson showed that genetic counselors often held widely divergent views as to the severity of conditions like XYY, hemophilia, or sickle-cell trait.

In spite of data that suggest that carriers are only occasionally at risk when under stress at high altitudes, or have minor pathological changes that show up in autopsies, the expected consensus that carriers for the sickle-cell gene are at no higher than normal risk of disability or death was not found.

A survey conducted in the early 1970s by a task force of the prestigious National Academy of Sciences revealed marked differences of opinion among physicians regarding the clinical significance of sickle-cell trait and the genetic basis of disease generally. Only about half the physicians polled recognized that sickle-cell trait was harmless and rarely caused problems. And different kinds of physicians showed marked disparities in judgment. The National Academy of Sciences survey showed that about 50 percent more pediatricians than family physicians accepted the innocuousness of the trait. On further questioning, only slightly more than half of family doctors clearly agreed that many human metabolic disorders are genetically determined—a fact correctly surmised by Sir Archibald Garrod in 1908.

Since this and other surveys reveal marked deficiencies in the knowledge that even physicians have about the nature and severity of genetic disease, it is all the more remarkable that public policy has begun to move so rapidly into the area of genetics.

Extraordinarily complex diseases like hypertension and diabetes have also been swept up in simplified genetic or familial models of causation, and mass screening drives have been instituted, with little or no attention to the intricate correlations of diet, stress and social forces in their actual genesis. Yet, a genetic model is unnecessary for estimating recurrence rates or establishing treatment regimens for either hypertension or diabetes. Here, as elsewhere, emphasis on genetic causation can deflect attention away from a critical cluster of environmental factors that may be crucially important in determining who gets sick and how we help them.

The Logic of Screening

Beyond the practical problems that can deflect such screening efforts as the sickle-cell program from the professed aim of preventive health care, there are basic epistemological problems with genetic screening as a way of knowing and healing. For example, on one level genetic screening proposes a change in emphasis from patient to *potential* patient, from disease to disease *propensity*. This shift can be actualized only where genetic screening provides a bona fide look into the future. Aside from the fact that virtually all genetic screening techniques rely on inferential data (one does not look at the gene but rather at the process that is thought to derive from it), the validity of screening is undercut by genetic theory itself. This theory stipulates that genetic characteristics—pathological or otherwise—become manifest only after interaction with some environmental factor and may, therefore, never become manifest at all. Observers who overlook this basic tenet end up with a distorted one-cause theory of development, and a distorted view of genetics. For example, one public-health researcher has remarked that genetic testing "has become a new form of public health, in which local health officers . . . are having to re-orient their thinking a little beyond the more usual search for bacteria or viruses, as they begin to seek out the abnormal genes."

What is incorrect in this view is its emphasis on "normal" and "abnormal." Genes are, certainly, the vehicle for species standardization and familial similarity and, in this sense, we may speak of "normal" genes. But genes are equally the basis of natural variety, uniqueness and change. Ultimately, many forms of neonatal screening seem to overlook this vital fact of the *normality* of genetic differences while, at the same time, asserting that genes count more than any other factor in determining physical well-being. This latter assumption is the topic of an enormous intellectual controversy of our age, and how it is to be resolved remains to be determined. In the interim, it is important to try to understand why the idea of genetic causation has taken such a powerful hold on the scientific and public mind alike.

Heredity Under Control

> Our basic ethical choice as we consider man's new control over himself, over his body and his mind as well as over his society and environment, is still what it was when primitive men holed up in caves and made fires. Chance versus control. Should we leave the fruits of human reproduction to take shape at random, keeping our children dependent on accidents of romance and genetic endowment, of sexual lottery or what one physician calls "the meiotic roulette of his parents' chromosomes?" Or should we be responsible about it, that is, exercise our rational and human choice, no longer submissively trusting to the blind worship of raw nature?
>
> —JOSEPH FLETCHER, *The Ethics of Genetic Control* (1974)

The revolution in modern genetics is occurring not only in laboratories and state-mandated testing programs, but also in the media, in government, and in foundation funding offices as well. Why has genetics suddenly taken on so much importance? Why have genetic programs been incorporated so quickly—and often with so little forethought—into massive public-health programs? A recent very loose survey of public opinion by a medium-size California newspaper suggests some interesting answers to these questions. "Genetic research offers a possible solution to hundreds of problems," one reader reported. "Engineering genetics is really producing a new controlled person," said another. A third reader exhorted biologists to "move ahead with every possible combination, to a positive future."

Taken together, these statements provide a summary of the kinds of expectations and beliefs about genetics that fuel our collective imagination. We believe that genetics is all-powerful; we believe that we have an imperative need to apply it; and we believe that it gives us an unprecedented degree of control over our destinies. The naïve faith in science, technology and progress reflected here reminds one of the public attitudes toward atomic power after World War II. In fact, genetics is only the most recent science to be acclaimed as the key to a "positive future." Since the seventeenth-century pronouncements of Francis Bacon on the virtues of the scientific method, scientific knowledge has been equated with power. In Bacon's writings, power over nature meant power to change nature for the betterment of man. The traditional Faustian view that scientific knowledge and power could be purchased only at some spiritual cost was ultimately countered by a series of "miraculous" discoveries, especially in medicine and physics. These discoveries not only validated the scientific method as an epistemology, but supported Bacon's (and, more recently, Claude Bernard and Walter Cannon's) definition of progress as the assertion of human control over nature—human and otherwise. The achievement of absolute control has eluded our grasp—but, more ominously, it undergirded the drive toward total power that we saw in post–World War I Germany.

Perhaps because humanity must daily confront the consequences of "chance biological events," the drive to suppress and control what Carl Jung called the "nuisance and danger" of chance events is nowhere more obvious than in the study of human biology and its applicability to human behavior.

The discovery of an information-transmitting molecule by Avery and his co-workers in the early 1940s brought genetics to the forefront in this battle against biological chance. Deoxyribonucleic acid ("DNA," as the molecule was later called) opened up not just the possibility of a new weapon in this war, but the possibility of a whole new arsenal. "This is something which has long been the dream of geneticists," wrote Oswald Avery, an early DNA pioneer. "Up until now the mutations [have been] . . . unpredictable and random and *chance* changes" (italics added). Avery's assessment of the magnitude of his discoveries was no exaggeration; they were truly momentous. But his high hopes for DNA's power to fully control the end points of complex biochemical reactions and ultimately to delimit biological events have so far not been realized.

It is possible that some of the delay in recognizing complexity in gene action has been conditioned by the language that scientists have used to describe the gene and its activities. Early choices of key descriptors of gene action suggest that genetic language was shaped by, and later reflected, the expectations that genes would provide a controlling function through a simple cause-and-effect link. These choices are clearly illustrated in the earliest papers of the Molecular Biology Era, where the genetic code is presented to us as a command language, directing cellular function through the intermediaries of "regulators," "codes," "translators," and "messengers." In the molecular world of the gene, it is anathema to refuse, disregard or disobey. In this sense, the genetic language adopted by the scientist is more akin to a military code than to a dialectic. Genes are said to "direct" protein synthesis, for instance, when in fact they provide merely the first template from which a whole host of highly coordinated, and other gene-mediated activities ensue to produce the specified protein product.*

This limiting emphasis is unfortunate, because the genetic code has more recently been shown to be more similar to real language, with its reliance on syntax and feedback, than anyone had previously recognized. The actual language of the genes is a medium for education and even inquiry, as well as command. The products of the gene can only be "read out" or understood in an environment that is conditioned by outside forces—and by the instructions given by other genes. To be deciphered, gene activities require adequate preparations in the cellular milieu. Functional gene products, for instance, often require enzymatic activation or additional components to be added for full activity. The precursor molecule of insulin is an example of the former; the *globin* portion of hemoglobin an example of the latter. Genes themselves often contain spliced-in pieces of meaningless segments, which have to be translated, recognized and excised before a functional code is produced.

Thus, the activities of any one gene require the participation and cooperation of others. Genes will work only in the right environment, and their impact may be reversed, dampened or enhanced by the conditions that surround their operation.

The reason this fundamental observation has been ignored may be psycho-cultural or simply part of the Western scientific drive to control nature. Whatever the cause, controlling the genetic material has proved insufficient to elucidate "higher" functions like sight, hearing and speech. The reasons for these failures may rest on the

* Tertiary changes, including the final molecular folding needed to achieve a 3-dimensional structure, may be entirely gene-independent.

limitations of the biological cause-and-effect model, or on the present limits of our ability to direct mutations in the DNA molecule, or, most pessimistically, on the existence of intrinsic limits to what control of DNA means for controlling the end points of our quest for domination: human qualities themselves.

Still, Avery's preoccupation with control persists, often in surprising quarters. The Nobel laureate geneticist Joshua Lederberg, for example, has proposed, perhaps rhetorically, that "if a superior individual . . . is identified, why not copy it directly, rather than suffer all the risks of recombinational disruption, including those of sex? . . . Leave sexual reproduction for experimental purposes; when a suitable type is ascertained, take care to maintain it by clonal propagation."

But even such genetic control would not assure control over the actual product, since nongenetic forces would undoubtedly supervene to shift the direction and end point of development far from the desired prototype. Similarly, Joseph Fletcher, a bioethicist, vows that "we cannot accept the 'invisible hand' of blind chance or random nature in genetics." For Fletcher, progress is a human imperative actively opposed by the random events of recombination and segregation in genetic systems. Taking a different view of the matter isn't just courting error, according to Fletcher, it's ethically wrong.

> Producing our children by sexual roulette without preconceptive and uterine control, simply taking pot luck, from random sexual combinations, is irresponsible—now that we can be genetically selective and know how to monitor against congenital infirmities. As we learn to direct mutations medically we should do so. Not to control when we can is immoral. This way it will be much easier to assure our children that

they really are here because they were wanted, that they were born on purpose.

What is perhaps most astonishing about Lederberg's and Fletcher's views is the ultimate irony of suppressing variation. Evolutionary *progress* (itself a subjective term) actually depends on the processes of chromosomal recombination—the very processes condemned here as reactionary. Although there are repair systems that minimize perpetuation of errors, genes are intrinsically unstable and susceptible to chance events; this, together with sexual reproduction and natural selection, is what makes evolution possible. Even in transcribing their coded messages, mistakes occur. Attempts to control natural genetic recombination (e.g., the Ptolemies' habit of brother-sister intermarriage to perpetuate their "divine" genes) tend to have paradoxical effects in increased incidences of genetically based pathologies as less and less variability is generated. Moreover, we cannot now "direct mutations medically," nor is there any expectation that we shall be able to do so in the future. It is possible, of course, to perform prenatal tests to "screen out" grossly defective embryos or fetuses through selective abortion; but this process seems to occur naturally on a far greater scale. Perhaps as many as 60–80 percent of all chromosomally abnormal embryos are spontaneously aborted.

One might assert that attempts to improve the naturally occurring genetic mechanisms that foster (and control) change are limited not only by the law of diminishing returns, but also by the far reaches of human hubris. Joseph Fletcher disagrees, asserting that control is a necessary component of human action. He writes, "Controlling the quality of life is not negative; it just rejects what fails to come up to a positive standard." The dangers of such an approach are so clear as to be self-evident: Who determines the "positive standard"? Who (and what) determines the qualities that we collectively find so burdensome as to require a policy of genetic control? A state policy? Should *all* developing fetuses with Down syndrome be aborted? And what about those with a chromosomal variant of unknown significance? And, in between, what of those one in every two thousand newborns who has an extra Y chromosome that may—but only *may*—predispose the person to problems in later life?

Each of these manifest dilemmas rests on the false premise that a common genetic or chromosomal constitution damns the carrier to a common fate. While it is true that a specified genetic composition can define broad boundaries of expression, the facts of the matter are quite clear: Even the most egregious chromosomal rearrangement or genetic effect is modified by environmental interaction, and by the actions of other genes. Children with Down syndrome, for instance, show a remarkable range of abilities—from almost total dependence (with IQ scores falling in the 20–40 range) to virtual self-sufficiency (with IQ scores as high as 80). And simple training of the mothers of Down syndrome infants dramatically improves the acquisition of skills over the "expected" values.

The classical physical stereotypes of Down syndrome children—notably a thick tongue, low-set ears, and a "simian" crease in the palm—are also found in other children. A systematic study of some seventeen physical stigmata previously presumed to be indicative of Down syndrome showed that a child with Down syndrome may have as few as five stigmata, and a "normal" infant as many as eleven. Again and again, we slip into dogmatic presumptions of causality, only to have nature toss up an exception. This lesson is lost and relearned almost constantly in genetics.

"Proofs" of simple genetic associations or causes for

disorders follow a similarly pock-marked trail. For instance, several researchers have presented seemingly "hard" evidence that individuals with certain blood groups are predisposed to certain behavioral norms, only to have their generalizations dashed by subsequent studies. Blood group O was reported in 1961 to be found preferentially in persons with manic depression, and presumably therefore to be causally associated with the disease. Seven years elapsed before this readily tested observation was refuted!

A more recent case in point concerns the possible association of susceptibility to lung cancer in smokers to the ability of their cells to activate an otherwise relatively harmless chemical to a carcinogen. In this case, the original report strongly concluded that a heritable, single-gene basis existed for the increased ability of cells in smokers with lung cancer to convert 3-methylcholanthrene to its active form. Later, at least one of the initial researchers asked industry to begin to consider including this test in its screening programs. The data have since been criticized because family studies were not done and a welter of conflicting reports have been published, demonstrating the actual uncertainty of the initial results.

Even the original animal studies which provided the experimental basis for the human extrapolations have begun to come unstuck. Whereas the initial data showed a single dominant allele demarcating susceptible strains of mice, and its recessive analog resistant ones, recent data have painted a picture of sharply contrasting results and inexplicable complexity. What was initially an orderly picture of rankings of mice strains for inductibility locked in step with the same rankings for susceptibility to formation of tumors, has been shaken by paradoxical findings which we will see again in Chapter 7.

A simple change of a single variable not only broke the ranks, but reversed the genetically predicted sequence of results! A team of researchers from the Jackson Laboratory in Bar Harbor, Maine, showed in 1979 that the rankings could be turned on their heads by simply using a lower dose of carcinogen.* Not only were these results unanticipated, but the nature of the differences was greatest for the mice with the most similar enzymatic activity, suggesting that much more than simple Mendelian genetics was necessary to explain animal susceptibility to cancer.

A report on a search for genetic basis for the predilection of persons with chronic obstructive lung disease to breathe less deeply and frequently than normal provides an example of more appropriate caution in presenting data. These authors concluded that in the absence of genetic markers on a large family that could be studied over many generations, it was not clear whether genetics or familial-environmental factors could explain the differences.

In spite of these and similar reservations, genetics continues to be used dogmatically to explain phenomena that are otherwise complex. One measure of the psychological compulsion to control is the extent of this arbitrariness, rigidity and dogmatism as a characteristic of the use of new data or ideas in any particular epistemology. Such an attitude was demonstrated by the early advocate of eugenics Albert Edward Wiggam, who stated that "nothing is more certain . . . than that godly parents beget godly children and ungodly stock spawns a godless brood." A review of genetic textbooks and other recent writings on genetics shows that dogmatism dogs the heels of genetics to this day.

In the first edition of *You and Heredity*† (a book

*L. M. Prehn and E. M. Lawler, "Rank Order of Sarcoma Susceptibility Among Mouse Strains Reverses with Low Concentrations of Carcinogen," *Science*, Vol. 204 (1979), 309–10.

†Amram Scheinfeld, *You and Heredity*, (Frederick A. Stokes Co., New York, 1939), p. 9.

acclaimed as the best popularization of genetic science in the 1940s), for example, readers were told:

> You believe the astronomer when he tells you that, on October 26th in the year 2144, at thirty-four minutes and twelve seconds past twelve o'clock noon there will be a total eclipse of the sun. You believe this because time and again the predictions have come true.
>
> You must now likewise prepare yourself to believe the geneticist when he tells you that a specific gene, whose presence can as yet be only deduced, will nevertheless at such and such a time do such and such things and create such and such effects. . . .

In the revised edition of the book, a bevy of scientific reviewers let the latter paragraphs pass without any more qualification than that they ought to be modified with such phrases as "under certain specified conditions," or that the geneticist must make "more reservations" than the astronomer, or that the gene action "is complicated by innumerable factors." Even with these caveats, the paragraph overstates the case, implying that some form of absolute knowledge was, at the time of the book's publication, within the grasp of geneticists.

This assertion suffers a bit in light of a statement made later on in the text that "every man and every woman at conception received 24 *[sic!]* chromosomes from each parent, or 48 in all." This error was perpetuated for seventeen more years, until Tijo and Levan demonstrated that human cells *characteristically* have 46 chromosomes, and significant variations from this norm are uncommon.

Review of the literature of the period reveals the powerful suasion of dogma, as one researcher after another failed to count the right number of chromosomes even as techniques improved. The "Emperor's Clothes" phenomenon persisted until two courageous researchers outside the grip of orthodoxy broke the news in 1956.

Contemporary genetic texts carry on this tradition of dogmatism. A disturbingly high proportion identify various traits as examples of single-gene controlled capacities when evidence to this effect is lacking. Tongue-rolling and the manner in which one folds one's hands (left thumb over right, or vice versa) are perhaps the most commonly used examples. In 1975 geneticist N. G. Martin reviewed almost a decade of previous studies on these traits and then conducted his own twin experiments to verify his hypothesis. He found no evidence whatsoever of a genetic component to these traits and concluded that "most of the variance in these traits arises from the specific environmental influences and chances that affect hand-clasping and tongue-rolling." In spite of this unequivocal report, more than 65 percent of the twenty-six laboratory manuals and genetic texts surveyed by Barnes and Mertens in 1976 reported tongue-rolling as genetic, and roughly 25 percent linked hand-clasping and genes.

In another study, Nicholson and Mertens reviewed the fifty-seven human characteristics used in twenty-six laboratory manuals as examples of single-gene–determined human traits. From a list including astigmatism, eye size, eyelash length, freckles, hair texture, lip shape, mongolian (*sic*) eye fold, nose shape and (of all things) singing voice, these investigators reported that seventeen of the traits did not appear in Victor McKusick's comprehensive and authoritative catalogue, *Mendelian Inheritance in Man*. Nicholson and Mertens concluded that the absence of these traits from McKusick's inventory "implies that they may have no genetic basis whatsoever, or that they may have a polygenic (complex) mode of inheritance."

Other sources of popular treatments of genetics perpetuate the myth that we have certainty about genetic causation. A recently published book on sociobiology *theory* insists on presenting hypothetical propositions as fact. Richard Dawkin's book, *The Selfish Gene,* presents evolutionary systems as if it were *known,* rather than hypothesized, that the major driving force in evolution is gene-perpetuating behavior. Similarly, a recent popular treatment on the nature-nurture controversy in the journal *Human Nature** states that "When Mozart composed his first minuets at the age of five, his precocious talent reflected his father's teaching, his opportunities to get to a harpsichord . . . and a little help from his genes." Humility would have us say that we simply don't know what help (if any) Mozart got.

Occasionally, popularizations of genetics err in treating genetic defects as if they were perfectly analogous to infectious disease. Victor McKusick points out the related pitfall of treating some infectious or teratogenic syndromes as if they were inherited. The intent of such accounts may be salutary (to increase our study of genetic origins of defects) but the exaggeration of the extent of genetic disease is not. For example, in the book *Know Your Genes,* the first statement that greets the reader (in capitals, no less) is: "YOU ARE A CARRIER OF FOUR TO EIGHT DEFECTIVE GENES!" As we have seen, one statistical basis for this statement is a small and specialized number of Japanese studies of recessive lethal genetic disorders. While it is likely that each of us does carry some indeterminate number of deleterious recessive genes, *no one* knows their actual number or nature, or whether they are found more frequently in some populations—or individuals—than in others.†

*Sandra Scarr and Richard A. Weinberg, "Attitudes, Interests and IQ," *Human Nature,* Vol. 1 (April 1978), pp. 29–36.

†Indeed, the awareness of deleterious genes (average 2.2/person) suggests that they would have a Poisson distribution in the population, meaning that many more persons would likely have *no* such genes than would have four to eight.

The trend toward sensationalized, dogmatic, cause-and-effect genetics is not limited to commercial or educational media. It can also be found in the literature of public health. For instance, the National Institutes of Health, in a recent publication entitled "What Are the Facts About Genetic Disease?" launched a barrage of overstatements about genetic conditions and their impact on health. The N.I.H. reported that "at least 40 percent" of all infant mortality is the result of genetic factors; that genetic defects are present in nearly 5 percent of all births; and that about one third of all patients admitted to pediatric wards are there for genetic reasons. In a similar vein, the staff of the House Subcommittee on Health in the Interstate Commerce Committee reported in 1976 that four fifths of all mental retardation was "genetically related."

None of these assertions is supported by an analysis of the data. The assertion that 40 percent of infant mortality can be attributed to genetic factors cannot be substantiated even if *all* congenital defects and respiratory-distress syndromes are presumed to be genetically based—a fact totally at odds with current medical understanding. Nor are "genetic diseases" characteristic of 5 percent of newborns, unless one is willing to call all congenital defects and chromosomal variations (including potentially ambiguous ones) genetic "diseases." The assumption that genetic disorders characterize one third of all pediatric inpatients rests on similar suppositions. In fact, however, most major congenital defects are the result of complex interactions between environmental and hereditary factors; many forms of cleft-lip and palate, heart, spinal-cord and neural-tube defects are examples. The commonest causes of childhood morbidity, namely respiratory disease, diarrhea or other gastrointestinal problems are often the result of prematurity and external factors with little, if any, genetic input.

Other public-health publications repeat and not infrequently magnify these distortions. For instance, the California Department of Education's 1977 publication "Genetic Conditions" contains all the above inaccuracies and adds some novel ones. This booklet asserts that "contrary to popular belief, genetic conditions and birth defects are common health problems."

The reported facts paint a different picture. The British Columbia Genetic Registry, widely recognized as one of the most thorough files on birth defects and genetic disorders, records less than 0.5 percent of Canadian births as having a demonstrable genetic abnormality (0.08 percent dominant, 0.11 percent recessive, 0.04 percent X-linked, and 0.2 percent chromosomal disorders). Even allowing for a portion of the 4.38 percent of the births in British Columbia that have congenital malformations to be largely genetically caused, the contribution of single gene defects to disability is astonishingly low. Only when multifactorial diseases are included, does the total number of conditions *directly or indirectly* attributable to genetic makeup approximate the popularly reported figures. And, as we shall see, this last category is the one that should be most suspect as reflecting "pure" gene contributions.

Similarly, the records on admissions to hospitals with active pediatric services reveal substantially fewer gene-related problems than popularly believed. For instance, the records of Johns Hopkins Hospital for 1963–69 revealed that single-gene defects accounted for 6.4 percent and chromosome abnormalities another 0.7 percent of all admissions.

The best estimates place no more than 12 percent of pediatric admissions in the category of single gene, chromosomal or multifactorial disease; an additional 18 percent of admissions are for somal congenital malformations.

The widely held view that such genetic disorders, while rare, are nonetheless catastrophically costly is also undergoing reexamination as more prophylactic approaches to treatment are tried out and more home therapies are instituted. In California, for instance, where there is a special California Children's Services program designed to care for children and young adults (under twenty-one) with genetic disease, the full amount requested in the 1979–80 budget for treating all hemoglobinopathies, cystic fibrosis, Huntington disease and hemophilia was 3.362 million dollars, a mere 1.18 percent of the total budget for public health in the state.

At least part of the reason for this confusion of facts and figures is the lumping together of single-gene and chromosome abnormalities with those whose origins are more complex. Most of the above cited errors stem from including the so-called "polygenic" or multifactorial diseases with the more simply inherited disorders. I suspect that this error stems from the tendency of many medical researchers and geneticists to believe that all the disorders that we now call "multifactorial" will in time be shown to be the composite of many separate, single-gene abnormalities.

Whether truth reveals this state of nature is still an open question, but in the meantime, we might do well to examine the roots of the presumption that natural systems lend themselves to strict causal analyses of genetic and environmental contributions.

7

Nature and Nurture Revisited

The interaction of nature and nurture is one of the central problems of genetics.

—J.B.S. HALDANE (CIRCA 1946)

It is remarkable . . . that so little systematic attention, either in research or in theory, has been paid to the implications, extent, and meaningful analysis of [nature-nurture] interactions.

—L. ERLENMEYER KIMLING (CIRCA 1972)

It makes sense only, perhaps, to divide the influences that affect us into two categories: those outside us and those inside. Bifurcating the world into these two domains simplifies our understanding. Since the 1890s, when the work of August Weismann demonstrated the isolation of the germ plasm from the effects of environment, we have come to think of the genes as a kind of cellular sanctum sanctorum, the inside of the inside, which is passed down, intact, from one generation to the next. Factors in the environment constitute an external set of determinants that is less fixed. In this view, as environmental forces constantly impinge on the organism, resonating with its internally set responses, they reinforce, dampen or activate genetic instructions, but they cannot change them. B. F. Skinner described this net of forces as they operate on an individual. "A person's behavior," he writes in *Beyond Freedom and Dignity*, "is determined by a genetic endowment traceable to the evolutionary history of the species and by the environmental circumstances to which as an individual he has been exposed."

At face value, this formulation appears faultless because it leaves nothing out. But in a way it does. Genes and environments do not simply "add up" to produce a whole. The manner in which nature and nurture interact to cause biological organisms to flourish or decline is an extraordinarily complex problem. Indeed, the way in which a given genetic makeup actually leads to the development of an integrated organism is considered by biologists and philosophers alike to be the central problem of genetics. Its roots can be traced to the origin of our thinking about the actions of genes—to the ten years that followed the simultaneous rediscovery of Mendel's principles by Correns, Tschermak, and de Vries in 1900.

In the early 1900s, Danish botanist Wilhelm Johannsen studied the effects of different stocks of bean plants grown under nearly identical conditions. Johannsen expected the bean seeds of "wild" plants to vary much more widely than those taken from an individual, inbred variety. As expected, he found that the wild beans generated plants with a wide range of second-generation bean weights and shapes. But to his astonishment, he also found that the inbred beans, harvested from a single plant, seemed to vary just as widely. Depending on such circumstantial (or environmental) factors as pod position or order of emergence, some beans were larger, some plumper, and others were rounder than genetically identical beans on the same plant. Johannsen planted these second-generation inbred beans and produced the results he had originally expected: *on the average,* the new plants produced from any bean of a given stock made beans that were almost identical in size, weight and color to those on the original parent plant. More importantly, each line of bean plants bred true for generation after generation. In a word that he coined in 1909, Johannsen distinguished the internal sources of control over the visible characteristics of the beans from the visible characteristics of the plants themselves. He called these external characteristics the plant's *phenotype,* and the aggregate of the genetic contribution, the *genotype.* In Johannsen's first classic formulation of the problem, then, bean phenotypes varied from a genetically controlled norm as a product of such environmental factors as sunlight and moisture.

This early bilateral view of scientific genetics gave the contemporary thinkers (circa 1910–1930) a simple way of conceptualizing the relative roles of nature and nurture. Some, like those who advocated eugenic reform, placed all the weight of causation on the genetic side of the ledger. To others, the very simplicity of the distinction was appealing. Albert Edward Wiggam, a noted eugenicist, defined the two causative factors in a 1924 book.*

> Two sets of factors are involved in the development of an individual, and doubtless the same two sets of factors are responsible for racial development, or evolution. One category of factors is intrinsic and seems to depend on the mechanisms that are involved in cell multiplication and differentiation: all such factors are here included under the term "heredity." The other category of factors is extrinsic and seems to involve both environment and training: these factors are usually included together under the term "environment."

*Albert Edward Wiggam, *The Fruit of the Family Tree* (Bobbs-Merrill, Indianapolis, 1924).

In making this partition, Wiggam made it clear that real social progress could be accomplished only through genetic reform.

While contemporary researchers reject the notion that either genes *or* environment can independently determine such a complex phenomenon as behavior, one still finds abundantly oversimplified models of environment-genotype interactions. For example, in studies of animal or insect behavior, a particular environmental event may be said to "release" a genetically based behavior. With the rare exception of the genes that control certain stereotypic behaviors where only a single environmental trigger is necessary to activate them (and here the larva-chamber cleaning that is performed by bees comes to mind), the expression of a particular gene-associated behavior usually occurs only as a culmination of a series of environmental interactions. Thus, a male rat will not express its genetically based propensity for aggressive behavior unless its early postnatal environment permits a surge of testosterone and it is exposed to a later set of environmental stimuli, including visual and olfactory clues, which culminate in releasing the behavior in the adult.

In spite of advances in our understanding, even the best present-day interactionist models appear insufficient to explain the complexity of forces that interact to *cause* the emergence of higher human properties or functions. As late as 1972, behavior geneticist L. Erlenmeyer Kimling, writing in *Genetics, Environment and Behavior,* decried the absence of any real progress in our understanding of gene-environment interactions of fifty years ago! It seems likely that human behaviors (and especially those we consider distinctively human) are the product of a very great many genotype-environmental interactions, each one steering the organism this way or that, but definitely leaving it in a place other than the place that it was in before the interaction. But whether or not even one such interaction is adequately described by simple models of causation is questionable.

Recent research suggests that all higher organisms are destined to emerge as distinctly different individuals, *irrespective* of their genetic makeup. This principle means that genetically identical single cells—or whole organisms like uniovular twins—will not be identical in their development or behavior. Indeterminate or random exposure to agents or substances in the environment can account for part of this variability.

One study involved comparative ranking of the aggressiveness of three strains of genetically identical laboratory mice. The three strains were rank-ordered in terms of their aggressiveness and activity levels by two teams of researchers. One team, led by behavior geneticist Benson Ginsburg, ranked the strains A, B and C. A second team, led by John Scott at Bar Harbor, Maine, ranked the same strains in exactly the reverse order—that is, C, B and A! Painstaking research finally identified the critical variable that produced these dramatically different outcomes: in Ginsburg's experiments, the mice had been transferred from cage to cage by hand; while John Scott's people used tongs in handling. Apparently, a lifetime of the extra stress produced by the tail-tweaking Bar Harbor procedure influenced the expression of the genes that indirectly determine aggressiveness in mice.

Similarly, a 1975 study tested the hypothesis that environmental conditions are really secondary to genetic factors. Taking their lead from the study cited above, groups in Bar Harbor and in Italy reared genetically identical mice under conditions as nearly alike as possible. The parameter on which these researchers focused their attention was the critical one of brain weight. If genes were, in fact, the critical determinants in higher nervous functions (so the argument went) little or no variation in brain weights would be found between genetically identical mice raised in different environments. The gene-directed effects, if strong enough, would be expected to override any minor perturbations coming from the environment.

These quite reasonable expectations were dashed. Of the eleven strains of mice tested, no common gene-based characteristics were found. The one that consistently had the largest brain weight in Italy was only seventh-largest in Maine. The relative loss in brain weight in the Bar Harbor strain could not be attributed to total body-weight differences, nor were other environmental differences found that were capable of explaining the brain differential. As a result of these investigations, Italian and American workers alike urged colleagues to use "extreme caution" in extrapolating data from one mouse strain to another—or even between genetically identical animals reared under slightly different conditions. And even genetically identical mice raised in near-identical conditions will still differ significantly in their response to a toxic or pathologic agent, like a cancer-causing chemical. Such results suggest "caution" is a misnomer when extrapolations are made to humans, whose environments differ radically.

This outcome suggests some of the difficulty underlying any attempt to determine the genetic contribution to health or disease processes in highly structured organisms. Asking a geneticist to specify this contribution in precise, predictable terms is roughly comparable to asking a historian to set out exactly what actions by what persons on a minute-by-minute basis led to the fall of Rome. Certainly Rome was *caused* to fall, but the factors involved may be said to be beyond our knowing, because of the time, space and distance involved.* Inevitably some information is lost and some is confounded by random events. Genetic systems, likewise, are parts of historical processes; and in all but the most exceptional cases (sickle-cell anemia is a notable example) it is impossible to assign a directly causative role of the genes to some higher characteristic.

Even a complete knowledge of the genes of an organism tells us little about it as a whole. Gunther Stent has pointed out the absurdity of believing that we could transmit a cat through interstellar space, simply by transmitting its DNA sequences in their entirety. Had we all the genetic information of an individual animal computerized and catalogued to the point that we knew every biological molecule, its relative amount and time of production, for any given sequence, we might not be any closer to knowing whether the genes we held belonged to an osprey or an otter. For instance, unless we know the developmental forces that lead keratin into osprey feathers rather than into otter fur, we can't predict which body covering we will get. And some of those forces undoubtedly lie outside the nuclear genes.

In genetics, as in geography, the map is not the landscape. Those additional nongenetic features that would allow us to construct all the intricate surface variables of each organism are still outside our grasp. And there are theoretical reasons for believing that they may remain so. A total knowledge of physics and chemistry does not in itself enable us to recognize a machine, much less the knowledge of biology of an organism. We have to know the plan that allows us to put the information together in time and place to make a whole. As Michael

* For the sake of accuracy, it should be mentioned that some historians, notably J. B. Bury, believed that chance alone could account for the fall of Rome.

Polanyi has pointed out, we require an understanding of the guiding principles that make a machine run in order to use the information meaningfully. A clockmaker cannot meaningfully use physical or chemical observations to explain the operation of clocks unless he understands the operational principles of a clock. Similarly, even if we had total knowledge of the genetic physiochemistry of our living subject, we would now be hard-pressed to explain the integrating forces that coordinate his functions.

This limitation is due not so much to a dearth of information as to an overwhelming array of confounding events. Thus, whether I enjoy a lifetime of physical well-being or suffer the constant pain of a particularly debilitating back condition may be *independent* of the fact that I carry the HLA B27 gene. Part of the uncertainty is the result of gene action being indirect as well as direct.

One theory is that B27 heightens the sensitivity of the carrier to certain bacterial antigens, and that in the ensuing immune reaction, the body's own constituents are attacked. As we saw in Chapter 3, while 95 percent of the persons suffering from ankylosing spondylitis carry this marker, a much smaller number of the carriers ever develop clinically significant symptoms. One reviewer has optimistically observed that knowing these risk factors can help to focus awareness on the kinds of diseases most likely to affect each individual during his lifetime. Well and good, but we should also be sensitive to the risk that pursuit of the genetic roots of such conditions as ankylosing spondylitis will bleed resources away from the search for the critical, presumably environmental, factors that trigger the disease.

The critical problem of modern genetic medicine is (or ought to be) the exploration of the delicate nexus between gene and environment. Genes do not, themselves, cause disease. They may, of course, prepare the individual organism to react to stimuli, or external inputs in special ways, but the particular disease process is always heavily influenced by the indirect action of unrelated genes; by the environment of the soma generally; by the environment in which the developing soma exists; and (as is most obvious with the human species) by the way the organism as a whole affects the surrounding environment. Thus, the symptoms of phenylketonuria (PKU) do not begin to develop until the fetus is exposed to an external source of phenylalanine. More tellingly, if such exposure is blocked postnatally, little or no disease process ensues. Ultimately such environmental effects feed back into and accelerate or retard genetic propensities.

This model of feedback loops and synergistic interaction holds for both the most complex and apparently simple phenomena comprehended by the new genetics. For example, there are currently some 160 specific single-gene defects in humans that are associated with a clear-cut risk of cancer. Breast cancer alone may include five hereditary types; and cancer of the colon as many as fourteen. But in aggregate, these associations likely account for less than 20 percent of the total cancer burden. Genetic contributions like these still fall far short of absolute causation. Each gene-directed miscue (e.g., in DNA repair after damage) requires an internal or external trigger, or insult, and most require some external event like sunlight exposure to produce the irreparable damage to the genetic material that appears to be the precursor of malignant change. Here again, not everyone with the gene develops cancer either.

Even where causation appears absolute, close inspection almost always reveals the exception. For instance, where a chromosome test on a newborn reveals a missing segment on the Number-13 chromosome, the child has a 95 percent chance of developing the sometimes fatal eye tumor known as retinoblastoma. On one level, this observation may be taken as a sufficient statement of cause—but, on another level, one ought properly to ask what caused the segment to break off in the first place; or, what about the 5 percent who go scot free? Here the answers are at least as likely to be environmental—and changeable—as they are genetic.

Despite the dramatic consistency with which a few scattered tumors are linked to genes, the problem of cancer as a whole is much more than a problem in genetics. Through my own work in cancer immunology, I learned that genetic models of causation are almost always gross simplifications of nature. Genetically "identical" mice again and again have been shown to respond quite differently in cancer studies. Such differences in susceptibility to carcinogenic effects of a given chemical have now been shown to originate in systematic, nongenetic events, and not simply from chance variation. These confounding events—such as birth order, intrauterine stress, and other as yet unknown factors—cannot currently be controlled by even the most prescient of experimenters.

Hypertension is a major health problem that may turn out to be a model of gene-environment interactions. Some researchers are convinced that at least in blacks, genes play a major role in its genesis. Many physicians believe that the high blood pressure that is virtually endemic among inner-city blacks may be due to the genetic background as well as the high-salt environment and such factors as stress and poverty. Animal models provide ample vindication to such a complex theory.

During the 1960s a strain of rats was isolated with a clear, heritable propensity to high blood pressure. Unless they are subjected to stress, however, rats of this sort will not develop hypertension. They seem rather to be unduly sensitive to environmental stimuli—disturbances as minor as the addition of a pinch of salt to their diets. Their sensitivity, moreover, is not confined to nutrients. When these rats were placed in a situation wherein they had to endure minor electric shock in order to feed, their blood pressure zoomed. Similarly, "psychogenic" stress is thought to be a factor in human hypertension, but how such stress operates to generate high blood pressure is still unknown. That stress *is* involved is now universally agreed. Is it a primary cause of hypertension, or a secondary, contributing one? What forms of hypertension are triggered by environmental forces? These and other questions remain unanswered.

These examples suggest on a small scale what I suspect is true of human knowledge in the aggregate—that we may speak of "influences" and "correlations" in areas of comparative certainty, and of "causes" and "effects" very rarely indeed. In genetics, we may say that a complex chain of relationships and reactions link genes to their final effects on the organism. Whether these links are random (in the sense of infinitely complex) or predictable cannot now be answered with certainty.

Since Wilhelm Johannsen's pioneering work, genetics has come a long way. From the simple division of genes and environment at the outset, we have passed through a second-generation model of gene-environment interaction and seem poised on the brink of a third great era in theory-building. Beyond the inadequacies of the nature-nurture, gene-environment interactionist model, we can glimpse a biological reality in which these poles simply don't exist. It is a world that can be seen only indirectly, through gene products one or more steps removed from their source, and then only after they have interacted with the environment. It is a world in which the very cells that make up our "insides" are playing host to such "outside" bodies as the mitochondria, each with its own genotype and evolutionary history. From the perspective of cellular biology and biochemistry, then, the distinction between nature and nurture, between "inside" and "out," makes little sense. And so it is that we will likely have increasing difficulty specifying where environment leaves off and genotype begins.

Our world of causation may thus be closer to that of the ancient Taoist philosophers, for whom nature and nurture were arbitrary categories. According to traditional texts, nature and nurture coexist in a "net of causation" *(Chi Kang)*. In this world, events and forces condition each other simultaneously, and linear chains of causation are relevant to an organism only to the extent that they affect one another in a collapsed interaction. For the Chinese, historical forces are not as important as are the momentary interactions that continually redefine each object's place in the world, and condition events to each other. To paraphrase Carl Jung, the problem of human understanding is not so much the necessity to know how events A, B, C, D and E led up to one another, as the need to know how it is that other events A′, B′, C′, D′ and E′ happen to coexist at the same time.

A finer look at genetic systems suggests that such a view may be more useful than has previously been allowed.

The Limits of Genetic Causation

The effect of a gene mutation of the phenotype is determined by the interaction of the mutant gene with all other genes and with the environment during epigenesis [development].
—C. H. WADDINGTON, *Towards a Theoretical Biology* (1969)

Cat fanciers, such as myself, are likely at some time in their careers to confront the riddle of the cat with six toes. Where does the extra toe come from and what does it mean? One theory has it that six toes are a sign of special favor; the cat so born will be graceful, bright and well loved. Others have taken six toes as a bad omen, a sign that prodigies of all kinds are about to be loosed on the world. As for the problem of where the extra toe comes from—or, for that matter, why five toes are the norm—since Mendel, one school has held that six-toed cats are that way because they bear a single dominant gene not carried by their compatriots. A closer look at the history of our understanding shows that this simplistic view has been muddied with complications. In 1934, Sewall Wright, a famous population geneticist, set out to solve the problem of *polydactyly* ("the condition of having many or more than the usual number of, toes or fingers") once and for all.

Wright studied guinea pigs. His first findings seemed to show that the tendency might be genetic, but that more than one gene was involved. Later he suspected that a single gene might be responsible, but that its ability "to penetrate" the animal's normal developmental pattern was deficient. Other findings seemed to show that the genetic effects could be distorted by a wide variety of nutritional and other environmental factors. Polydactyly in guinea pigs, according to Wright, was actually a product of the influence of several different genes, woven through complex developmental pathways and threaded in and out of the environment in a very complex and intricate pattern.*

Wright's work on the problem of extra toes represents a leap forward in modern genetic thought. While his results have been refined in terms of our minute understanding of some of the agencies involved, his conclusions remain essentially valid. They are a model of well-tempered and comprehensive scientific investigation—the kind of investigation that we need in conducting a critical review of the welter of data produced by modern genetic inquiry. Since Wright, the accumulated evidence indicates that genetic causation is buffered by a plethora of small agencies and events affecting the expression of the gene. Some of these effects are generalized and stable, some are not.

For instance, the relatively common congenital malformation known as *spina bifida* (from the Latin for "split spine") is one of a class of neural-tube defects that seem to have a genetic component. People who survive the defect

*Such "multifactorial causation" is, of course, not new in its literal sense: in the mid-nineteenth century, the celebrated French surgeon Ambroise Paré listed thirteen causes of congenital malformations, including the devil, so that parents could ascribe their misfortune to any of several external forces.

are more likely to produce children afflicted with the problem than people who don't have it; and, even where the parents don't express the abnormality themselves, the birth of one affected infant increases the risk that the next child will have the defect too. Other studies show that where both mother or father had had spina bifida, fully 3 percent—or thirty times the expected number—of their children had it too. Spina bifida, like Wright's polydactyly, is a product of polygenic, or multifactorial, inheritance; rarely, it appears to be caused by environmental factors independent of genetic predisposition. The authors of virtually every definitive study of spina bifida cases found it impossible to separate out all the confounding genetic and nongenetic factors that cause the disease.

What factors muddy the hoped-for clarity of genetic explanations and rob them of their power to explain biological phenomena? For one thing, as Sewall Wright suggested, gene action is frequently conditioned by the impact of other genes carried by the organism. For example, investigators of human diabetes have recently found that several genes are involved in the production of the precursors to susceptibility to this multifactorial disease. And where a specific gene product is not manufactured, some persons are up to 1,400 times more likely to develop juvenile-onset diabetes mellitus than a brother or sister who does produce it (HLA B8).

In an attempt to preserve some of the clarity of single-gene explanations in a multiple-cause world, Wright proposed a theory of critical genetic "thresholds." Genes with *strong penetrance,* according to this view, show through all but the most unusual circumstances. *Low-penetrance* genes, by contrast, may manifest their effects only when environmental conditions enable them to do so, and then probably only when whole banks of genes are operating.

Most dominantly inherited conditions in humans, while single-gene–caused, span a spectrum of penetrance.

Thus, because of as yet unknown intervening variables, only a small percentage of persons with the genetic makeup for the dominantly inherited disease called *neurofibromatosis* exhibit its full-blown characteristics. The dense fibromas that cause tissue displacement and damage (particularly in nerves entering the brain) may be totally absent in a known carrier. And in only about 1 in 10 affected persons will a tumor become malignant. All that may show are some minor skeletal changes or a striking café-au-lait–colored spot on the skin, a spot that may appear in any number of persons without the neurofibromatosis capability.

Perhaps most unsettling of all, the *origin* of the neurofibromatosis gene appears to make a difference. When the gene comes from the mother, the condition appears to be consistently more severe than when the indistinguishable gene comes from the father. This apparent paradox is partly resolved by recognizing that the mother also differs from the father as the place where development occurs.

Thus, a second factor that can complicate genetic causation is the impact of the uterine environment on the developing organism. While we are predisposed to think of the embryo as leading a life removed from the stresses and corruptions of postnatal existence, new evidence suggests that the real picture may be quite different. While endowed with some genetically fixed functions, most embryonic qualities and processes are highly plastic and open to environmental shaping. In place of the old womb-as-sanctuary idea, it might be more accurate to think of the uterus as a kind of negotiating room, where the exact combination of internal (genetic) and external (environmental) forces that determine the phenotype is worked out. Neither gene nor uterine environment determines the outcome; but the interaction between the two is, in many cases, decisive.

Stress responsiveness serves as an instance in the extraordinary complexity of intrauterine forces that can shape an animal's repertoire of behavioral forces. The genetic makeup that somehow controls stress resistance in an animal that develops in the uterus of one mother will work entirely differently in the uterus of another. Fostering, nurturing and rearing studies in which an animal of a given makeup is transferred to the uterus of another while still an embryo, show that the behavioral response of a particular strain of mouse to electric shock or other stresses is heavily conditioned by the factors present in its prenatal uterine environment. In turn, stressful events during pregnancy appear to alter the balance of maternal hormones from the adrenal gland in a way that can disrupt the normal development of the fetus's own adrenal glands. Behavioral geneticists have consistently shown that newborn mice or rats that have been subjected to such prenatal environmental conditions behave differently in terms of activity levels, emotionality, and general response to stress itself than do their normally reared peers.

In some cases, genes seem to "determine" the phenotype of the organism only insofar as they lay down basic susceptibilities to environmental agencies. Here, the interactional causes of a particular structure or process lean toward the environmental. And in some cases the uterine environment seems to have an overriding curative effect on embryonic vulnerability. One experiment with mice, for example, demonstrated that cortisone "caused" cleft palate in half the offspring of a particular strain injected with the drug. But when embryos from this susceptible strain were transplanted to the uterus of a mother known to be resistant to cortisone's effects, the birth defect appeared at a radically lower frequency postnatally. For as yet unknown reasons, a similar "maternal," uterus-mediated effect on cleft palate has not yet been found in humans; yet, when a parent has a cleft palate with (or without) a cleft lip, an affected child will usually express the same pattern.

Similarly, one strain of mice called "A/Jax" is characterized by irregularities in the number of vertebrae; yet embryos transplanted into the uterus of a non-type-A mother very rarely show the irregularities at birth.

By the same token, the uterine environment can be dramatically affected by substances ingested by the mother. Perhaps the most infamous example of this fact is the thalidomide-induced birth defects widely reported on in the late 1950s and early 1960s. Thalidomide produced two broad classes of embryological defects, depending on the time of its ingestion in relation to a critical point between the twenty-second and the twenty-third day of gestation. The first consisted of muscle paralysis, absence of ears, faulty development of the base of the spine, and disordered kidney development. The second class of defects included a closed anus, hernia, and a skeletal malformation of the thumb. What was not widely reported at the time was that these drug-induced defects went for some time unnoticed because they were rare, and possibly because they resembled no fewer than seven naturally occurring, and apparently genetically based, defects. In England and West Germany, this resemblance, together with bureaucratic delays, forestalled recognition of thalidomide's causative role in producing the more general and devastating defects of the arms (phocomelia) which became its hallmark.

One of the modern genetic developments that permit the trial of genetic-environmental hypotheses is the existence of genetically uniform strains of mice. Produced by repeated brother-sister matings, these mice are, for all practical purposes, genetically identical. It was with some surprise, therefore, that modern teratologists (birth-defect experts) found that even such identical mice, when exposed to the same environmental stimuli, reacted quite differently. For instance, genetically identical offspring of mothers who have been exposed to a teratogenic drug can show a complete range of responses, from apparently normal development to gross malformation even in the same litter, depending in part on the physical location in the uterus.

More baffling still is the discovery that the embryos of some teratogen-injected mothers can demonstrate an astonishing ability to recover from the drug's effects. In one experiment, reported in the April 1975 issue of the *Journal of Experimental Zoology,** pregnant female mice were injected with a teratogenic substance (trypan blue); gross deformities appeared in 95 percent of the embryos by the following day. But when researchers allowed a week to pass between injection and dissection, no more than 20 percent of the embryos were found to be developing abnormally. Somehow, eight of nine damaged embryos had recovered from the environmental trauma.

Other "genetically determined" characteristics also turn out to be heavily influenced by as yet unknown factors—and even when a large array of these factors are tested for their contribution to "congenital" defects, no clear-cut picture emerges. For example, researchers once systematically examined the combined effects of maternal weight, seasonal gestation, embryonic position in the uterus, fetal weight, and several other factors, on the development of spontaneous (as compared to the cor-

*Max Hamburgh, Mark Ehrlich, and Gail Nathanson, "Some Additional Observations Relating to the Mechanism of Trypan Blue Induced Teratogenesis," *Journal of Experimental Zoology,* Vol. 192 (1975), pp. 1–11. First observations were made at the egg cylinder stage; the second, after implantation when organ development was well underway.

tisone-induced variety just discussed) cleft palate in mice. After years of work, these researchers concluded that various nongenetic factors must be skewing their results in unpredictable ways.

Other scientists have isolated a few of these non-genetic factors (such as maternal stress and diet) in the genesis of cleft palate; but the specific mechanisms by which these factors affect the embryo remain baffling. In humans, cleft palate seems almost always to defy analysis, since even affected parents only rarely have affected children. Yet, having once had a defective child, the risk of it recurring is in the order of 5 percent, irrespective of the parent's condition.

As with other disease processes and malformations, this impression of weak genetic causation may result from an underestimation of the number of possible genes that cause the defect. Also, either environment or genetic makeup may sometimes account for the appearance of the same malformation; more often, however, the cause is likely to lie in a highly complex interaction between the gene and its surroundings.

As to the value of statistical approximations of genetic contributions to the observed variability in a polygenic trait, in the words of a correspondent writing in the March 9, 1973 *Nature*, "there can be no argument that statistical models are only a substitute for a more basic understanding."

For any given polygenic defect, there are thus a multitude of causes. The malformation rates for these conditions are, in fact, slightly increased in near relatives, especially brothers and sisters of affected individuals, suggesting common genetic pathways. Similarly, malformation rates in the United States are intermediate between those of countries that contributed substantially to our population, such as England, and those that contributed less, such as Sweden. These facts, coupled with the observation that malformation rates increase with inbreeding, led one researcher to suggest a genetic hypothesis—but a complex one.

Population geneticist James Neel proposed that many congenital malformations might be partly explained by the following model: From the population's gene pool, some organisms receive a disproportionately large allotment of defective genes. In these affected organisms, development takes place without adequate "buffering"* or other normalizing events, and the embryo is laid bare to noxious environmental forces, some of which disrupt it sufficiently to cause defects. Other researchers, notably Newton Morton, believe that the evidence is far from conclusive and, in fact, militates against a genetic hypothesis. The fact that persons of Japanese and Caucasian ancestry acquire the same relative incidence of birth defects of the spinal cord (spina bifida) once they have been residents of Hawaii, while the native indexes of Japan and the United States differ dramatically, led Morton to conclude that forces in the native environments, and not the genes, are largely responsible for spinal-cord defects.

Very subtle variations in the composition of a gene can be used to explain why certain people present the classic symptomatology of a genetic disease, while others do not. Such genetic heterogeneity is the rule for recessively inherited conditions, and this has been proved in the case of phenylketonuria, Tay-Sachs disease, galactosemia and virtually every other sufficiently studied single-gene disability. In cystic fibrosis, it may be that slight differences in the genetic makeup can explain cases where none of the

*The word *buffering* was first used by British embryologist C. H. Waddington to describe the way slight changes in the genotype could be absorbed or corrected by the developing organism without its deviating from a particular developmental pathway.

pancreatic defects that are virtually universal concomitants of the disease appear. But are such variations products of genetic variation or environmental modification, or some other factors?

In some instances disease mechanisms seem to be activated or modified by the intervention of some random event—for example, the chance contraction of infectious hepatitis or measles—that activates or modifies the individual's propensity. In other cases, we simply don't know how to explain the variety of individual reactions. Microbiological processes seem to be shaped and sometimes ruled by such imponderables as a person's "physical constitution" or "character style."

A recent issue of the *Journal of the American Medical Association,* however, reports the case of a remarkable man who had completed combat training at several Special Forces schools, was successfully performing his job duties, and was jogging eight miles a day. At age thirty this man had sought medical help because of some difficulty breathing while running. The diagnosis: cystic fibrosis. His doctors described his previous adaptation as remarkable—a highly compensated genetic defect. Certainly this man is exceptional; but exceptions exist for virtually all the classic genetic diseases, debunking the fatalistic notion of uniform prognosis in single-gene disease.

Thus, the widely held expectation that death is a virtual certainty for any teenager with cystic fibrosis will have to be revised upward. Behavior geneticists in particular recognize that the actual effects of any gene depend on its context. This context includes: the genetic background, the population in which it is being expressed, the cultural setting, and the prevailing environment. All of these components interact to dampen, reinforce or suppress the potential effects of the genes.

Lesch-Nyhan disease is an example of the ways in

which genetics can affect the organism's environment—and vice versa. Lesch-Nyhan is generally considered to be a result of a deficiency in an enzyme produced by a recessive, sex-linked gene. Affected males are mentally retarded and suffer from some degree of spasticity. Other signs of the disease are compulsive biting of the fingers, arms, lips and other parts of the body—symptoms that begin to appear at about two years of age. Dr. Nyhan, the co-discoverer of the enzymatic basis for this metabolic defect, has referred to the syndrome as an example of how genes alone can determine a behavior in humans.

Other researchers—notably Drs. Dancis, Alpert, Anderson and Herrman—see things differently. In their case studies at Columbia University of children afflicted with Lesch-Nyhan disease, these observers found that the presence of parents, and the specifics of the setting affected the incidence of self-destructive behavior; they also found that the self-destructive behaviors responded to suggestion techniques. Was behavioral conditioning affecting ex post facto the genetic constitution? Or was the problem rather with Nyhan's model of genetic determinism?

In a brilliant gamble, Dancis and his co-workers tested the idea that children "try out" all behaviors, including painful ones, on a random schedule. Whereas normal children try out and then reject self-destructive behavior, Dancis reasoned that Lesch-Nyhan children may simply get locked into a particularly unfortunate pattern, because they cannot process pain meaningfully. Thus, the problem may not be one of genetic predisposition to head-banging, but rather a genetically based failure of adaptive feedback mechanisms. The crucial difference between the two views, of course, is that in the case of a genetically programmed behavior, one would not put much hope in environmental remediation; while in the case of a genetically mediated deficiency that merely *contributed* to aberrant behavior, one might hope for change. In favor of that latter hypothesis, Dancis found that the two boys participating in the experiment did respond to selective reenforcement and were able to suppress self-destructive behavior to a remarkable degree.

If single-gene–based conditions can show such variation, what of the behaviors determined by more than one gene and strongly influenced by the social context? Aggression is one such behavior that has lately attracted the attention of behavior geneticists. In their original studies of the problem, Benson Ginsburg and his associates seem to show a clear-cut association between specific male Y chromosomes and aggressive behavior in mice. This dramatic finding quickly came to the attention of other researchers, whose interest was not so much in mouse as in man, and not so much genetic or scientific as political. A writer in *Fortune* magazine, for instance, interpreted Ginsburg's work as proving the existence of an aggression-determining genotype. Later work, however, did not support the Ginsburg group's original hypotheses. To the surprise of the research team, experiments seemed to show that at least two gene groups *not* on the Y chromosome were involved in the expression of aggression. As with Wright's polydactylic guinea pigs, the pattern here was one of multiple genes functioning as parts of a microscopic ecological system. Moreover, the multiple-gene effects appear to be modified in turn by other gene-directed systems, and so on, until the notion of "gene-caused aggression" is attenuated to virtual extinction.

In addition to the largely physical environmental factors discussed in the last chapter, social factors have been shown to influence dramatically the expression of aggression. A husband and wife team from Scandinavia, for instance, has demonstrated that three virtually independent social mechanisms, in addition to genetic factors, shape the aggressive behavior of experimental animals. Taking normally aggressive mice as their starting point, these two researchers demonstrated that otherwise combative animals could be rendered nonaggressive by any of the following social procedures: 1) allowing an otherwise aggressive animal to be defeated in combat; 2) handling the animal extensively in its early life; and 3) grouping them with similarly aggressive animals for long periods of time. Among the primates, aggression seems even more plastic. Experiments have shown that highly aggressive animals, when transplanted into a new social group, may cringe and supplicate where in more familiar surroundings they once raged. These effects persisted even when the brain's so-called "aggression centers" were electronically stimulated in the transplanted primate to reinforce the previous pattern.

Researchers have also shown that the experimental animal's early postnatal experience can exert a profound and potentially overriding effect on its aggressive behavior. Undernutrition early in life heightens the tendency toward aggressiveness in mice. Housing a male mouse with a rat will dampen its aggressive tendencies, as will drug treatments that prevent the normal surge of testosterone during the prepubertal period; conversely, some strains of mice show no effect at all of socialization or even total isolation from other mice on fighting behavior.

A careful review of the effects of early experience on other aggression-related behavioral parameters, such as the interval between provocation and fighting, avoidance conditioning, and defecation, showed not only variability, but the same reverse effects of early experience of the behavior between specific strains; for instance, the same handling procedure may dampen aggression in one strain and heighten it in another. Unlike the work of Ginsburg and Scott, where rankings of unmodified adult aggressive behavior varied for the same strain, other studies have shown that it is possible to predict behavior of inbred strains of mice, but only for animals *as a group*. It is still difficult to predict the behavior of a single mouse.

In spite of such data indicating the overwhelming pattern of genetic determination to be one of complexity, biologists like E. O. Wilson insist on adhering to the hypothesis that, to use his words, "aggressive responses vary according to the situation in a genetically programmed manner."

Clarity may be forthcoming, but at present the situation for a general field theory for genetic programs undergirding aggressive responses of individuals is bleak indeed. At a minimum, these experiments suggest that complex events, and not genes alone, act in concert to "determine" the likelihood of a particular aggressive predilection.

Alcoholism is another complicated behavioral syndrome that has attracted the attention of many behavior geneticists. This interest is based in part on the demographic findings that alcohol is metabolized differently by different ethnic groups, and that these groups show different patterns of alcohol abuse. On the average persons of Asian origin, for example, "clear" alcohol more slowly from their bloodstream than do Caucasians. And adopted children whose biological parents drank appear to show a predilection for alcohol even where drinking was not encouraged in their adoptive homes.

Support for the hypothesis that the predilection to drink might be genetic also comes from animal research. One line of animal experimentation has shown that certain strains of mice express a definite preference for alcohol, while others do not. The two strains most often studied have almost opposite tendencies. The black strain C57BL/6 exhibits a strong preference for alcohol solutions from 1 to 10 percent by volume; while the tan strain DBA/2 will avoid alcoholic drinks even when dehydrated to the point of exhaustion.

The clarity and simplicity of these findings have not, however, endured closer inquiry. Behavior geneticists working with the alcohol preferring black strains had their first setback when it was found that different genes seemed to "control" alcohol preference in different strains. A second, and more serious, obstacle to the genetic hypothesis appeared in the early 1970s. First, tan embryos were found to lose their teetotaling propensity, simply by rearing them in a C57BL/6 mother's uterus or allowing them to suckle her milk. Then, researchers found if they placed young "naïve" black mice with nondrinking adult tans; and, as a control, housed naïve tans with black C57BL/6's who had an already-established drinking habit, *both* groups of naïve mice acquired the behavioral characteristics of their older compatriots. The young black mice (who might otherwise have been heavy imbibers) became virtual teetotalers when reared with nondrinking tans. The tans, in turn, became drinkers after seven weeks of exposure. While neither strain assumed the total behavioral repertoire of their cohabiting role models, or ever really drank as much as or as little as the "pure" opposite strain, this experiment showed in a remarkable way how powerful social factors can be in influencing behavior. When analyzed in detail, less than half the observed differences in alcohol consumption among mouse strains could be attributed to genetic differences.

If verified for other strain combinations, this model would suggest that genetically similar mice housed together could come to interact in ways that allow the slight predilection of a single animal to spread through the group. A genetic substratum for the behavior would have to be present in only trace amounts to permit social amplification of this kind. In humans, this socially mediated learning is, as Theodosius Dobzhansky repeatedly emphasized, at least the equal of genetic shaping.

Two other leading population geneticists have recently identified the extremely long period of infant dependency as providing the critical challenges and conditions of development. They emphasize that a "purely 'cultural' inheritance" exists that, "in the case of parent-offspring interaction, is almost completely confounded with biological inheritance."

In all of these examples of complex causation, attempting to partition the genetic contribution is like trying to understand how a symphony evokes a particular feeling by studying the individual instrumental parts.

Cutting Through Genetic Determinism

There is no absolute knowledge. And those who lay claim to it, whether they are scientists or dogmatists, open the door to tragedy.
—JACOB BRONOWSKI, *The Ascent of Man* (1973)

At the root of the conviction that genes in fact determine behavior is the philosophical premise that everything is determined by a knowable sequence of causes. It is not, strictly speaking, just that every phenomenon is *caused;* but rather that particular factors, acting in particular sequences, inevitably produce particular results. That is, given the same sequence of factors acting on the same environment, one always gets the same predictable result. Over and against this view of things is the idea that identical factors, even when they appear in identical sequences, may produce different results. In an earlier age, proponents argued for one of these theories or the other—often couched in terms of determinism or free will—in their pure forms. But today, the problem is not so much one of choosing between these views as somehow explaining how it is that both can be true.

As a science, genetics was the child of nineteenth-century determinism. Mendel and other geneticists developed genetic ideas as a means of explaining the baffling variety and complexity of life in simple cause-and-effect terms. "One character, one gene" became the byword soon after the rediscovery of Mendel's work in 1900. Beadle and Tatum's 1940s work showing a "one gene, one enzyme" connection in the bread mold *Neurospora* seemed to be a logical extension of this early work. The simple linearity of these concepts led to the belief that complex phenomena could similarly be traced unerringly through chains of gene-caused events to their origins. By 1938, researchers had already hypothesized that the cause of each primary mental ability would be found to be distant and independent events, rooted in the genes. In the 1940s and 1950s, for instance, a provocative series of experiments seemed to validate the idea that there were innately determined, and hence, gene-directed, releasing mechanisms behind complicated animal behaviors; that genetic constitution determined such complex syndromes as epilepsy; that the basis for alcoholism lay in the genes; and that the genes controlled the very construction of the central nervous system. In the 1960s, ethologists published observations seeming to show that a whole spectrum of behaviors—from aggression to social altruism to courting—were delimited by genetic mechanisms. By the mid-1970s sociobiologists had extended the conclusions of ethology to the human condition itself.

The findings and implications of many of these studies were both fascinating and (by their analogy to man) alarming. One experiment on releasing mechanisms is illustrative of a whole genre, all of which point sharply to innate factors as the penultimate dictates of behavior.

The experiment in question employed pairs of kittens and rats that had been raised together from birth in a state of benign coexistence. This innocent relationship, however, could be radically disrupted by the intrusion of minuscule electric currents transposed to deep centers in the kittens' brains. By implanting an electrode in part of the *hippocampus* (one of the brain centers that mediate aggression), the experimenter could, seemingly at will, direct an extraordinarily complex and coordinated sequence of behaviors.

In one graphic experiment, the administration of a tiny electric current resulted in a stereotypic attack-and-kill response that had never before been experienced or observed by the kitten in question. Upon administering the current, the previously playful and retiring kitten stiffened, crouched and lowered its tail, tip twitching. Without hesitation, it began a slow, purposeful stalking of the unsuspecting rat. With a perfectly calculated leap, and two or three carefully placed bites, the kitten broke the rat's spine.

No cat-and-mouse playfulness was visible here, but rather a seemingly preprogrammed atavistic killing. Behavioral geneticists dubbed such behavior "silent stalking" and published studies showing that even docile feline strains that previously never evidenced this behavior and could not be taught it, nevertheless responded instantly to electrical stimulation as if a switch had been thrown in their brains.

The moral of these studies for genetic determinists was simple: aggression is genetically programmed. The species-specific, stereotypic nature of the behavior makes the conclusion appear inescapable: same behavior, same nervous system, same genes. But the complex interplay of visual and perhaps olfactory cues that trigger the behavior—after all, it "works" only when the appropriate target is present—bespeaks a more complicated story.

The simplest argument is that the maturation of the

nervous system must be intrinsically programmed in order to generate such universal, stereotyped behavioral responses. A corollary of this hypothesis is that since the genes are irremediable fixtures, any behavior that emanates from them in such a stereotyped and consistent pattern must also be innate and fixed. The implications for such a posture are awesome and already being felt at the level of state-mandated programs for treatment of the mentally disordered violent offender. In California in particular, special sentencing and treatment programs are being proposed as the only way to treat intractable and presumably ingrained violent tendencies among habitual criminals, precisely because of the currency of the belief in innate aggressiveness.

But the fundamental assumption of this hypothesis and the programs predicated on its validity is that there is an irrevocable commitment toward certain highly coordinated patterns of violent behavior once an anatomical locus of violence is disordered. And the validity of this assumption rests on the premise that a kind of diagram of the circuit of the nervous system is encoded within the genes—a premise that is posited most notably by the neurobiologists Hubel and Wiesel and Victor Hamburger in their studies on perception.

At least three competing hypotheses have been advanced over the years to explain neural circuitry and its precision. In addition to Hamburger, Wiesel and Hubel's initial conviction that each neuron must have a genetically encoded program which directs its development and eventual link-ups with target organs, two other proposals have been put forth. The second, championed by Guenther Stent, is that much of the specificity in the nervous system arises out of trial and error, with growing nerve fibers which make "right" connections preferentially sustained over those which make "wrong" ones. The third hypothesis is really a synthesis of the first two. It posits that a *general* program of nerve circuitry is indeed gene-coded, but that a complex interplay of genetic and environmental forces is necessary to shape the final system. To understand the resistance to this latter, synthetic view, it is worth reviewing the seemingly compelling early data which led to the notion of genetic predetermination.

Early studies of the development of the nervous system seemed to support just this idea—that much, if not all, neurological development takes place along preordained pathways, regardless of the environmental circumstances of development. The eyes of a kitten, for instance, are "prewired" to the brain in a sense that 80 percent of the receptors in the visual area of the brain (the *striate cortex*) will be stimulated by light entering either eye, as soon after birth as the eyelids open. Other researchers, working from the behavioral end of the problem, demonstrated that whole behavior repertoires in some organisms seem to be heritable. Single genes apparently can determine much of the behavioral regimen of such bacteria as the common intestinal bacterium *Eschereschia coli,* and flies such as *Drosophila melanogaster* (fruit fly). *E. coli's* ability to move toward or away from certain chemical substances in the outside environment appears to be under genetic control. Similarly, Seymour Benzer was among the first to show that fruit flies *(Drosophila)* carry individual genes that determine their response to light or gravity. A good portion of the mating repertoire of fruit flies is similarly directed. Thus, a gene mutation that gives a yellow coloration to fruit flies also makes males of the same strain poor suitors.

Struck by both the neuroembryonic and behavioral evidence, behavior geneticists have sought gene-behavior links in higher organisms. One such inquiry has concerned a mutant mouse that reels about like a man with six or seven martinis under his belt. The problems of this "reeler mouse" turn out to be caused by an atypical development of the connections between the nerves in the eye and those in the brain. In normal mice, the eye neurones begin to sprout and stretch out, eventually connecting with the inside of that laminated sandwich of brain cells called the cortex. In the reeler, however, the first nerve projections stop short of the necessary cortical connection, and each successive wave of neural projections backs up behind this first, misdirected one. Besides being "wrong" anatomically, the reeler's cell connections are more loosely packed and randomly arrayed than are normal ones.

Up to this point, the reeler's development makes a good case for proponents of genetic determinism, and for those who believe genes can determine behavior. Yet, even after this disastrously bad start, all is not lost. Somehow—and no one knows exactly how—the nerve cells that must connect with the early projections send out sprouts that eventually make correct neural connections. Where these specialized projections fail, the vagrant cell itself may still send out a sprout to link up with the exact axone necessary to make some sense of the input. Often these sprouts meander tortuously through many cell layers before making their proper connections.

Even with these heroic attempts to redress the problem, the reeler still reels; but the behavioral effect is minuscule compared to what an uncorrected, purely genetic model would predict. A mouse carrying genes that cause damage of the reeler magnitude ought to be totally incapable of receiving, organizing, or coordinating sensory inputs at all. It is as if a television set were mistakenly wired into a telephone outlet, and then, on its own, ferreted out an appropriate wall plug, sprouted a new set of connecting wires, and tapped into the juice. It seems a bit beside the point in such circumstances to complain that the picture isn't perfect.

One "deterministic" theory that has been advanced to explain the reeler's phenomenal recovery powers is that many nerves possess a kind of homing instinct so that each cell is genetically programmed to connect with its appropriate correspondent. Neurobiologists have theorized the existence of "wiring diagrams" that guide the outgrowth of nerves (and later blood vessels) to their target organs. Almost simultaneously, researchers have discovered biologically active substances that apparently can direct nerve formation to "targeted" end organs, thus helping to save some credibility for the deterministic thesis.

The most intensively studied of such substances, known as "nerve growth factor" or NGF for short, lacks the specificity sought by the early determinists for a molecule which would be organ and nerve specific. Instead, NGF is released by many different organs, and serves as a general homing device to draw nerve fibers to their vicinity.

This theory leaves unexplained how it is that the ostensibly built-in plans for neural circuits sometimes can compensate for errors in placement of neural projections and sometimes cannot, even within a single species. For example, the "chinchilla" gene predisposes Siamese cats to develop a reversed visual-nerve circuitry. Normally, half the neural projections from each eye go to the opposite side of the brain, permitting stereoscopic vision. (See Figure 3.)

In many, but not all, Siamese with the gene, the overlap is imperfect, leading to some optic-nerve fibers reaching the "wrong" side of the brain. Compensatory wiring schemas seem to minimize the functional effect of the genetic miscue; and perhaps more certainly, adaptations

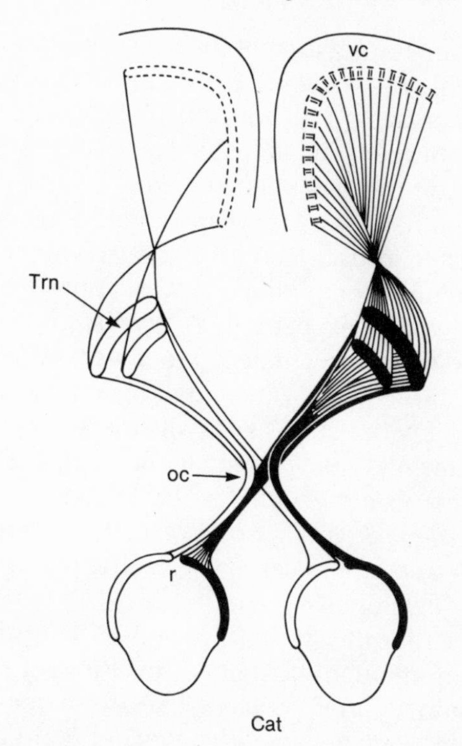

Figure 3.

The Route of Visual Images in the Eye of the Cat

This illustration schematically depicts the pathways that bring images that impinge on the retinas of a binocular animal into its visual cortex. Note that the images that strike the inner side of the left eye's retina are identical with those that strike the outer side of the opposite eye. Crossing of the pathways at the optic chiasma allows the one image from both eyes to be portrayed in the same area of the brain's visual cortex. This phenomenon, coupled with selective activation of "background" and "foreground" cells, is responsible for the cat's ability to see things in depth.

Used with permission from J. D. Pettigrew and M. Koniski, "Neurons Selective for Orientation and Binocular Disparity in the Visual Wulst of the Barn Owl *(Tyto alba)*," *Science*, Vol. 193 (1976), pp. 675–78.

in behavior and orientation of the eyes (recall that Siamese are almost always cross-eyed), compensate for the misplaced connections. Full binocular vision is nonetheless unlikely to develop as long as both eyes fail to focus consistently on the same visual field.

Are there other forces, perhaps external ones, that also exert a variable impact along the way? If the visual defect in all chinchilla-gene animals were purely genetically directed, we would expect a more consistent picture. As it is, different degrees of defect are routinely observed in different Siamese cats—while other species that also have the chinchilla gene (such as the Burmese and Persian cat families) show no visual disorientation at all and the albino rabbit makes no correction for its defective innervation. Obviously, then, environmental or developmental factors can distort—and others can correct—the possible harmful effects of a "deleterious" gene.

An acid test for the strength of genetic controls is to see how the information contained in the genes actually applies to a structural problem like the assembly of the neural connections between the eye and brain of a higher mammal. We now know that genes play only a provisional and very early role in determining the neural connections in the more primitive autonomic system. But many researchers have presumed that the innervation of the optic area of the brain's cortex had to be more precisely controlled because of the precision required in vision.

The nerve network that makes up the eye-brain link is incredibly complex, involving some 10^{10} individual nerve cells tied to specific end points in the brain. These eye-brain connections must be made with an extraordinary degree of precision in order that the whole system will function as an integrated whole. The primary reason for this extraordinary equipment is the exigencies imposed on the system by binocular vision. In species like the cat, owl and human, two eyes face front, resulting in a complete overlap of visual fields. In practice this means that each pinpoint of light is recorded twice, once in each eye. In turn, the electric signal from a cell discharging in the retina of one eye must somehow be routed to exactly the same cell in the visual cortex of the brain triggered by the other eye in order to form a single, fused image. (See Figure 3.) Without such precision we would all see double!

In order to obtain this degree of precision, the unavoidable errors and uncertainty involved in any finely detailed wiring diagram would have to be eliminated. One hypothesis, described most eloquently by Gunther Stent, is that the development of these components of the nervous system can be governed only by the roughest of "blueprints," and that the precision is accomplished by a kind of fine tuning based on later inputs.

Stent champions the idea that this fine tuning is wrought by an overproduction, during embryonic development, of cells and connections, among which only the "right" ones are somehow selected. According to this model, each cell in the brain is at first (i.e., at birth) connected to light receptors in the retina over a much wider area than is compatible with sharp vision. Light beginning to pour into the eye and, in turn, to those visual cortex cells in the brain, seems to touch off a process of selective reinforcement—and/or attrition—by which only those connections that record the same images reported by left and right eyes are left intact, much as a marble sculptor hews an image from a more amorphous whole. The importance of this realization is not lost on pediatricians, who must deal with disturbances in the early orientation and dominance of the eyes. *Amblyopia,* (muscular weakness in one eye) or *strabismus* (cross-eye orientation) or even *astigmatism* (deformations in the lens) can result in the wrong cell connections being reinforced, so that even when the early problem is corrected, fine vision may be lost in the subdominant eye—or in the astigmatic eye, which previously failed to focus properly.

Moreover, the "homing instinct" theory does not explain the nervous system's compensatory powers in cases wherein functioning is artificially distorted *after* initial development is complete. To examine the impact of postnatal distortions, Colin Blakemore, a British neurobiologist, surgically rotated a kitten's eye a quarter of a turn, replaced it, and then trained the kitten to recognize vertical stripes with its "good" eye only. Yet the kitten could still discriminate between vertically and horizontally striped doors with its "bad" eye only, and could act on these discriminations effectively. This result suggests a remarkable rewiring in the surgically treated eye's connections to the brain. For the kitten *not* to see the vertically striped door as horizontal, its nerves would have to form new connections which would reorient their visual world close to 90 degrees!

If animals can learn to overcome such extreme experimenter-induced distortions, it comes as no surprise that they can also compensate for some of the defects produced by genes. The current consensus of many researchers in this complex area is that genes may indeed prepare cells within the brain to anticipate vertical and horizontal lines or edges in space, but *the construction of a complete picture of the world* (one populated by oblique as well as right angles) *requires experience.*

For example, if kittens are fitted with goggles that permit one eye to see only horizontal stripes, and the other eye only vertical ones, those neurones with the capacity to respond to diagonal lines actually disappear. This kind of evidence strongly suggests that experience can shape the nerve circuitry of the visual field in animals. But do such effects apply to humans? Perhaps. Cree Indians, who are reared in a world of many diagonal lines (they live in teepees) are able to visually discern fine differences between lines placed along a diagonal with greater precision than those of us reared in a world of right-angle images. And persons whose astigmatism has misfocused the visual world in either the horizontal or the vertical dimension since birth, never regain the fine vision in those orientations, even when eyeglasses correct the defect.

Genetic explanations for perceptual skills still abound in the face of confounding evidence. Some researchers have even proposed that birds carry a kind of genetic "star map" by which to navigate by the night sky during their migrations. The indigo bunting seems to "know" what the stellar orientations mean at a given time of year, at least enough to orient its multithousand-mile flight. But as it turns out, experience *is* essential. The first exposure to the starry sky, for instance, is critical in laying down the navigational responses the adult bird will later have.

Proponents of the deterministic view seem to be right, then, only insofar as innate (presumably genetic) factors create the "negative" on the retinal screen upon which a photographic image may later impinge, and thereby indirectly create the conditions which permit the development of neural projections from the eyes to the brain. But the gene-directed rules that govern these adaptations may be of the most elegantly simple sort. If research on distant organisms like the South African clawed toad can be generalized to humans, it may be that the instructions for distribution of neural connections are as simple as "spread out to occupy all available sites," only later to be fine-tuned by experience. In turn, these neural connections can, in a sense, predetermine some of our perception of the physical world. But, equally clearly, genetic or experimental distortions of this world are compensated for by experience and as yet unknown mechanisms that undercut the gene's effect. *It is this uncertainty that puts the fatal cramp in determinism and makes a mockery of simple genetic causality* for higher functions, like intelligence, that must rely on perception for their realization.

What is true of the relatively automatic functions of perception is true as well of the genetic contribution to many so-called "instinctive" behaviors. Newly hatched ducklings, for example, respond to a special call from their mother by running to her and vocalizing loudly. The ducklings' listening, recognition and response unfold at critical times and without any discernible outside input—seemingly, that is, with all the hallmarks of a genetically based behavioral regimen. But a good detective wouldn't accept this explanation without more investigation: witness the following recently overheard dialogue:

WATSON. Here are the eggs you requested, Mr. Holmes. About two days to hatching time by my calculations.

HOLMES. Very good, Dr. Watson, Now tell me, what do you observe?

WATSON. Why nothing, sir. The ducks are still in their eggs.

HOLMES. Use all of your senses, Dr. Watson, and report your observations over the next hour.

WATSON *(under his breath)*. What bloody nonsense this is.

(One hour later.)

HOLMES. Well, Dr. Watson?

WATSON. They moved about a bit, sir.

HOLMES. Anything out of the ordinary?

WATSON. Nothing 'cept some cheeping sounds, sir.

HOLMES. Those cheeping sounds, where did they come from?

WATSON. Why, from inside the eggs, sir. Is that unusual?

HOLMES. Isn't it curious, Dr. Watson, that a duck should cheep before it is hatched?

WATSON. Why no, sir, why shouldn't it?

HOLMES. Well, Dr. Watson, the cheeping may be to bring the mother back to the nest, or keep the brood together, but note that in nature, the mother never leaves the nest during the last forty-eight hours, and the little ones will not travel about in their oval porcelain coaches.

WATSON. That reminds me, I've heard that the mother talks to her eggs sometimes.

HOLMES. Indeed she does, Watson.

WATSON. Don't you think that curious, Holmes?

HOLMES. No, Watson, it is only curious if the ducks inside cannot hear her.

WATSON *(surprised)*. What's that, sir?!

HOLMES. Well, then the hen would be talking to herself, wouldn't she?

WATSON. All right, all right. Let's assume that they can hear. What difference does that make?

HOLMES. You forget, Watson, we are trying to ascertain how ducks learn to recognize their mother's call.

WATSON. I'm afraid I don't understand, Holmes. Perhaps you would be good enough to explain.

HOLMES. Here are the two kinds of calls made by ducks. *(He picks up a wooden duck call off the desk.)* Compare the mother's very unique high-frequency contact-contentment call with that of her ducklings—while they're in their eggs. Remarkably similar, isn't it Watson?

WATSON. Yes, sir, but still—

HOLMES. Never mind, Watson. Let me go on. Suppose we were to open one of these shells and immobilize the vocal chords of the duckling with a drop of glue—ever so meticulously, mind you. What do you think would happen?

WATSON. Well, it wouldn't be able to cheep, would it?

HOLMES. That's not the point, Watson. Now we'll do it. *(Holding a pair of fine scissors, he takes an egg, opens the air space at the blunt end, lets the little head peep out, and gently holds it while he snips a tiny opening around the larynx and just touches the exposed neck with a drop of resin.)* There, that's done. Now put this egg away, out of earshot of the others. And bring all the ducks to me when they've hatched, but put a similar drop of glue on their vocal chords just before you bring them in so I can't tell them apart.

(The next day.)

WATSON. Here they are, sir, fit and chipper as can be.

HOLMES. No differences in behavior among them, Dr. Watson? No way I could tell my muted duckling from the others?

WATSON. No, absolutely not, sir. They are all alike as peas in a pod.

HOLMES. Very well then, bring them over here so they may hear my duck call.

WATSON. What's that for, Mr. Holmes?

HOLMES. You'll see, Watson. Please leave the room.

(Watson leaves.)

HOLMES *(loudly)*. Watson, come here, I need you!

(Watson rushes back in, breathless.)

WATSON. Holmes, are you all right?

HOLMES. Certainly, don't be foolish, Watson. Now, is this your duck? *(Holmes holds one fuzzy little ball forward.)*

WATSON *(in astonishment)*. Why yes, sir. I marked its toes the day of the operation. I was sure you would never tell it apart. How did you deduce it, sir?

HOLMES. It was really quite simple, my dear Watson. You see, the mother duck has a brooding call with frequency bands at 200, 800, 1600 and 2300 kilohertz. The duckling mimics the higher bands. Other avians like the chicken use the middle bands only. Now, if the duckling is to be able to recognize its own mother, it may very well need to subject *itself* to some practice before hatching—are you with me so far, Watson?

WATSON. Yes, yes, Holmes; do go on.

HOLMES. Well, I merely played the high-frequency mother-duck call to all of the newly hatched ducklings on my high-fidelity duck call. All but the one which didn't have the opportunity to hear itself came scurrying over to me, their surrogate mother. When I played a chicken's call, none of them came, showing that the self-imposed training is, in fact, specific for the high frequency unique to the duck. Obviously, if the newly hatched duck is to be able to respond quickly to its mother's call, something in its preemergent life must prepare it. Right, Watson? So its genes merely allow it to talk to itself.

WATSON. Holmes, you constantly amaze me!

HOLMES. Elementary, my dear Watson.

If Holmes had wanted to, he could have developed his solution further. The ducklings' responsiveness to noise actually develops throughout its life in the egg. Early in embryonic development it is sensitive to low-frequency sounds, and the ability for selective hearing of high frequency emerges only just before hatching. Deprived of these critical frequency inputs, the duckling is unable to pick out the calls of its particular species. Apparently, then, a critical period of exposure to sounds from various sources—including the duckling itself, its nest mates in neighboring eggs, or its mother—is essential for the appearance of this "innate" behavior in the neonate duckling. It may be that early sound input spurs the creation of, and later selectively sharpens, the duckling's ability to discriminate between specific sounds.

What of the duckling that was deprived of some of this input by Holmes's scissors? Interestingly enough, two days after hatching, devocalized ducklings acquire the ability to hear in the high-frequency range, but this ability wanes quickly if not reinforced again. It is as if a window briefly reopens shortly after hatching to allow the rich sound inputs of the natural world to wash over and stimulate the hearing apparatus. Genes do not prepare the duckling's ear to hear in isolation, but rather coordinate the sequence of neurological maturation according to reinforcement and conditioning schedules. Any genetic determination of high-frequency-sound perception is thus subject to environmental modification—or extinction.

A duckling that misses its auditory cues, an eye that loses its visual stimulation, or a behavioral region in the brain that fails to be activated by testosterone—all of these events result in the loss of what was otherwise a genetically predictable attribute. A duck *is* genetically programmed to selectively discern its mother's voice; kittens are genetically programmed to be able to see both horizontal *and* vertical edges; and males are genetically programmed to have neurons in their brains' sexual centers which can be activated by testosterone. But for all this genetic determinism, a chance event, or an abrupt environmental change, or a sudden surge in the wrong hormone at the wrong time can upset the apple cart of causation irrevocably, skewing our best predictions across the landscape. And, by analogy, some of this must be true for humans.

A new model of development is thus emerging from recent genetic experiments; it makes sense of the surprising, seemingly contradictory evidence produced during the 1950s and '60s. In this model, development is seen as a joint venture of genetic, chemical, functional and experiential factors. At the most basic level, the genes help fix fundamental structure and process. Very rarely, however, do genetic programs conform to a simple, deterministic, cause-and-effect format. Rather, the genes create the rudiments of physiological, perceptual and behavioral regimens, which themselves then open up or close off opportunities for environmental interaction. In the case of the visual processes we have been discussing in this chapter, the genes lay down receptive potentials that, when actualized by external stimulation, turn out to have a transforming effect on the original structures in the eye. Similarly, brain cells are genetically programmed to "see" certain visual stimuli and not others, thereby reinforcing or altering functional operation at the level of the eye. In this regard, all later perception may be as much contingent on the nature of the perceived world as on the genes that make perception possible.

For all these reasons, genes and environments can be expected to interact in ways that can confound analysis of their separate contributions. Experience filters into the development of an organism at critical periods when the genes have left open certain windows of indeterminacy. Outside stimulation—in the form of nutrients, electromagnetic and other forms of radiation, and even social interactions—stream through these windows and become integral parts of a changing interior environment. Nutritional deprivation during development can directly reduce the total number of cells in the brain's cortex, and other neurons appear to languish and die for want of adequate visual, auditory or general stimulation. Effects of environment or social impoverishment, then, can be expected to show up in the early biochemical makeup of the brain. And so they do, for humans as well as for rats.

This hypothetical model suggests the vastness of some

of the complexity likely to be involved in the genetic determination of later behavior. The genetic system may control the rate of proliferation within the brain, but the formation of *functional* pathways that link different levels with their sensory inputs and the reinforcement of cellular growth are almost certainly in large part the result of experience.

By allowing the outside world in during critical phases of development, the nature of the receptivity of an animal to its sensory world is thus encouraged, while not being overdetermined. The resulting plasticity increases the chances for a newly born organism to anticipate novel environmental changes that it would never have perceived if it were genetically locked into one mode of perception.

Thus, in all organisms genes seem to help fix the fundamentals of structure and process, insuring that at least minimum features of the environment can be perceived and acted upon without prior experience. But depending on an organism's specific evolution and adaptation, the balance between genetically determined and what we might call *genetically indeterminate* events will vary radically.

According to this evolutionary model, the more primitive an organism is, in a phylogenetic sense, the more likely it is to rely on its genetic program for anticipating the world. The fruit fly is thus an example of a highly genetically determined organism. From its emergence as an adult fly, it must meet extraordinary environmental demands with little or no time to learn which responses work best. Only the narrowest conditions of temperature and humidity will suffice for its survival at any one stage of development. Behavioral conditions are no less stringent. A whole constellation of tighly woven male-female interactions must take place if mating is to be successful, eggs have to be laid in a fermenting substratum, and skills in identifying food sources must be letter-perfect for the newborn to survive. All these activities require extraordinary coordination of complex behaviors. If all of them had to be learned, the fruit flies would simply not survive. Hence the genes fix its pattern of behavior, its inclinations, and even its choice of mates.

The fruit fly's world is not man's, however, any more than the total anonymity and rapaciousness that characterize the lives of flies is characteristic of the human condition. A long period of dependency during human infancy allows the previous generation of adults to introduce the child to the cultural artifacts of their world. Any genetically fixed responses would tend to close out the possibility of human interaction with the environment, thereby reducing the ways in which experience can enrich and transform the structures laid down by the genes. Human development is an edifice with so many "windows" that it is reductive and usually simply wrong to think of it in deterministic terms. Perhaps a previously unemphasized yet singular quality in human genetics that deserves more attention is its essential indeterminacy; and it is in this sense of programmed indeterminacy that the opposing views of genetics and environment discussed at the beginning of this chapter may be reconciled.

10

From Clockwork to Quantum

> We are searching for the human biogram . . . only when the machinery can be torn down on paper at the level of the cell and put together again will the properties of emotion and ethical judgment come clear.
> —E. O. Wilson, *Sociobiology, The New Synthesis* (1975)

In what was perhaps a last rush of Victorian optimism, some early-twentieth-century geneticists and developmental biologists held that discovering the ultimate causes of biological events was simply a matter of time. Wilhelm Roux championed this view, proclaiming that biology could achieve a comprehensive view of the natural world just as physics had done before it. Roux argued that it ought to be possible to take apart an embryo and reassemble it, much as one might do with a clock. His *Roux Archiv für Entwicklungsmechanik des Organismen* (circa 1902) (Roux Archives for Developmental Mechanics of the Organism) expressed the spirit of the times: Development started at one end and moved forward in a straight line, adding structure and complexity along the way.

Roux prophesied that biology would one day be characterized by the sort of deterministic surety that marked the physical sciences in the nineteenth century. This analogy was proved correct—but in a backward sort of way. The physics of Roux's day, that is, was shifting its attention toward a new energetics, one far subtler and more complex than anything dreamed of by the Enlightenment. In this new world of micro and macro events, there were no concrete and predictable exchanges between mass and vectoral force. Rather, the question had become: Where does force leave off and mass begin? In the new mechanics, parts were defined only through their interactions; and interactions between parts constantly shifted. As Werner Heisenberg, a brilliant young twentieth-century physicist observed, it was impossible to get the parts to hold still for scrutiny. The very energy necessary to visualize a particle on a hypothetical X-ray microscope screen would drive that particle away from the most steadfast observer's eye. And this uncertainty, Heisenberg was later to report, was *intrinsic* to the nature of the electron itself.

In a parallel development in biology, pioneering nineteenth-century German embryologists such as Hans Driesch and Oskar Hertwig quite appropriately rejected the mechanization of biological systems. Driesch, for one, was convinced that there were ordering forces in living things that defied the clockwork analogy. He hoped to test his ideas against those of Roux and Weismann, who believed that embryogenesis was simply the unfolding of developmental capacities from a storehouse of genetic information. Weismann had already generated powerful support for his model of the partitioning of the *germ plasm* (the sex cell line) showing that when a four-cell snail embryo was divided into two, its later development was irrevocably disrupted. Driesch, however, was able to demonstrate that this outcome was not universal. A single cell, split off from the multiplicity of cells in a sea-urchin embryo, could still form a complete, though diminutive, adult. "Is it possible to imagine," asked Driesch, "[that] a complex machine, unsymmetrical in three planes of space, could be divided hundreds and hundreds of times and still remain intact?"* Even the engineers in biology had to agree that the concretely mechanistic model was no longer adequate.

We now know that development is possible only because each cell in the embryo is more than one isolated genetic unit. It is a member of a tightly balanced and coordinated whole, such that, if one cell dies or drops out early in development, it can often be replaced by others, which assume that cell's identity. Further, the larger and more intricate the integration of this collective whole, the more the immediate environment of any individual cell is under control. But how do these control systems develop?

Some contemporary geneticists such as Jacques Monod of the Institut Pasteur or Bruce Alperts, of Princeton, maintain that all the information for performing crucial control maneuvers lies in the genes. And so far as bacteria are concerned, this additive genetic explanation seems to work. Alperts believes that control of development resides in the gene-directed elaboration of key molecules. He and other researchers are seeking a *morphogen*—a key hypothetical substance that cues specific developmental patterns or events.

Where higher organisms are involved, however, the additive view seems far too simplistic. Some genes are not in the nucleus at all; others are heavily controlled by intranuclear or even extracellular events. In multicellular organizations, ecologically determined effects are the rule, not the exception. To give but one example, the formation of the wing of a fruit fly requires at least thirty different

*This quote is taken from Driesch's famous treatise, "Experiments on the Egg of the Sea Urchin" *The Science and Philosophy of the Organism,* Vol. I (1908), 59ff (London, A. and C. Black).

genes acting in concert. Holistically inclined researchers such as Louis Wolpert believe that development of this sort is best understood by uncovering the *principles* behind it.

Wolpert and his co-workers see the development of organ systems with three-dimensional structural integrity as proceeding from "organizing gradients" rather than as a summation of many isolated gene-directed events. Presumably, no less genetically determined but irreducible forces operate along the body of a developing embryo, say around the periphery of the "bud" from which a new limb will form. Indeed, an organizing region called "PZ" has been found in the wing buds of chicken embryos; this region somehow sets the "polarity" and handedness of the whole structure. Similarly, Seilern-Aspang, an Austrian researcher, posits that the *morphogenetic* (literally, "form-generating") forces set up during regeneration are strong enough to control the growth of cancer. Tumor cells that are otherwise unresponsive to the forces that surround them in adult tissues, Seilern-Aspang reasons, may very well respond to the force fields generated during regeneration. Indeed, when he placed melanoma tumor cells into a regenerating tail of a newt, they appeared to fall into place and were carried along into the newly forming regenerate.

Even more astonishing than Seilern-Aspang's controversial and little-appreciated work are the studies of Beatrice Mintz, senior staff scientist at the Institute for Cancer Research, in Fox Chase, Pennsylvania, and Leroy Stevens of the Jackson Laboratory in Bar Harbor, Maine (both of whom I worked with as a student researcher). Injecting embryos with an otherwise fatal dose of Stevens' *teratoma* cells and injecting the resulting *blastocyst* into a suitably prepared uterus, Mintz was able to show that the tumors were incorporated into the normal tissues of the embryo, resuming their appropriate function and role!

Apparently, the tumor cells responded to homing cues and fields of force that are as yet only dimly understood by the most prescient minds in research.

Mintz's view is that malignancy (at least in the instance of teratomas) constitutes a response to a disorganized cellular environment rather than a mutational or genetic event. If the cells that are destined to become cancerous are returned to a sufficiently well-organized milieu (in this case, the cavity formed by a normal embryo in its early stages of development), their normality is reestablished.

Mintz also pioneered the development of a technique for fusing two genetically different fertilized eggs. When stripped of their outer coat and allowed to commingle, the embryo's cells combine into a superembryo, share their developmental instructions, and give rise to an intact mouse—one with four parents! [See Figure 4] In spite of an enormous genetic difference in parental traits (in some cases one pair of parents may even bear lethal genes), normal-appearing embryos survived and developed into adulthood. The apparent lesson from Mintz's work: Genetically disparate cells of the same species can read-in meaning and synthesize an integrated whole from unlike parts.

This work and others have contributed to a new awareness of subtlety and complexity in the consequences of gene action. The holistic view, that development is influenced or guided by (in Driesch's phrase) an "equipotential harmonious system," is but one of the theories challenging the conventional additive view of genetic determinism. Roux, in a simpler era, confidently prophesied the appearance of a Newton in biology; of a scientist who would be capable of resolving biologic contradictions and bringing to the life sciences the sort of deterministic surety which characterized nineteenth-century physics.

In the physical sciences, the complexity of natural

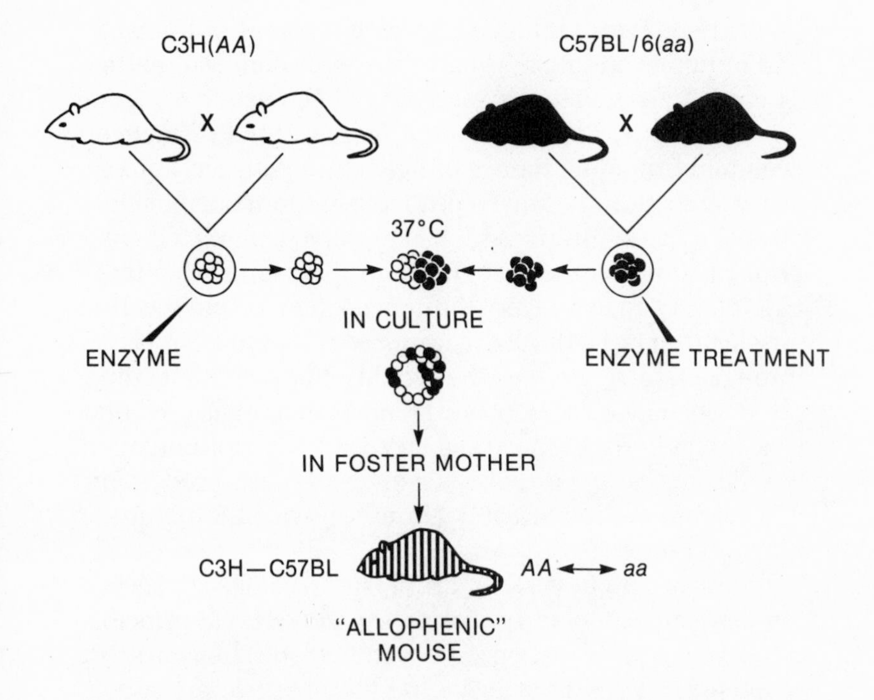

Experimental Production of Allophenic Mice by the Methods of Mintz.

The example shows two cleavage-stage embryos derived from the fertilized eggs of a pair of C3H and a pair of C57BL/6 inbred-strain parents. The enveloping *zona pellucida* of each explanted embryo is lysed in pronase and the embryos are aggregated by incubation at 37°C and cultured for a day. The resultant composite double-size blastocyst is then surgically transferred to the uterus of a pseudopregnant recipient previously mated with a sterile male. Embryo size regulation occurs soon after implantation, and development continues normally to birth. If both cell strains are adequately represented in the coat of the C3H⟷C57BL/6 allophenic animal, a pattern of fine transverse bands, representing the component *agouti (A/A)* and *nonagouti (a/a)* hair follicle clones, is seen. Other tissues, including the germ line, may also comprise both cell strains. Each cell, except for skeletal myoblasts, retains its individuality.

Adapted from Beatrice Mintz, "Gene Control of Mammalian Differentiation," *Annual Review of Genetics,* Vol. 8 (1974), pp. 411–70.

phenomena forced reluctant accommodations to ambiguity and multiple causation as early as the turn of this century. The old Cartesian one-cause-one-effect models of Bacon and Newton were shown to be insufficient in Einstein's account of the dual nature of light. And with Heisenberg, uncertainty itself became a principle in the epistemology of physics. Jacob Bronowski has observed that this new principle brought the absolutists up short. The Principle of Uncertainty discouraged those who had presumed that physical science afforded a window with limitless clarity through which to see the world. For the first time, physicists had to confront the fact that their ability to know was limited. In Bronowski's words, with Heisenberg it became necessary to acknowledge, however grudgingly, that "errors are inextricably bound up with the nature of human knowledge."

Today, we might wish for an Einstein or a Heisenberg; a thinker capable of formulating in genetics a principle comparable to Einstein's recognition of the bivalent character of light (that it can be both particle and wave, matter and energy). That is, if Wolpert is right about the impact of synergistic design effects on development, we must somehow account for the interactions between these architectural forces and the clockwork effects of individual genes, and then again between these architectures and the environment. It may be that Driesch was right, and biology will follow physics into ideas of causation that one might, in everyday life, categorize as paranormal. In any event, it is surely an error to think of "blueprints" in genes (a term coined in a recent Public Broadcast System program on the genetic revolution) as causative in the old Newtonian sense of the term. With the advent of computer-simulated embryonic development, as refined by the Levinthals at Columbia University, it is just possible that embryologists will shift away from an insistence on strict, linear causality.

In addition to the problems of understanding interactional systems, there are particular problems in the finite "mechanics" of genetic units—for example, How shall we begin to account for the myriad forces that can affect a change in the structure or impact of the individual gene?

We now should have reason to believe that observable measurement of a human attribute like blood pressure may result from any one of an infinity of gene-environment interactions. Can the "noise" of random events disrupt our ability to recognize the genetic contribution to any reading? Our certainty that the genetics inputs to a blood pressure reading, as with estimating electron positions, may be minimal. With blood pressure, there is no exact reading, much less a set input of the genes for any given individual.

The original notion of the "normal distribution" formed the basis of Heisenberg's uncertainty principle: when gazing at faraway stars, it was not simply the average of different observations which told "where" any individual star "was" in space, but rather the aggregate of the observations which described the *probability* of where they could be found. For the stars, the bell-shaped curve was an approximation; for the electron, the described probability function *was* reality.

The limiting factor in this instance was the reality of the electron, not our inability to measure finely enough. Continuously varying characters of human beings may be quite similar with regard to pinpointing the genetic contributions to their reality, as random events and forces set the outcome of initially tightly controlled genetic events adrift.

It may prove to be that genetic changes like mutations are analogous to radioactive decay, where the average "risk" of a radium atom disintegrating can be measured, while the *moment* at which such an event occurs in an individual atom remains indeterminate. As with radiation, a mutational event is "caused," but the frame shift here is an unpredictable one—a quantum jump in potential activity of the gene, not a gradual one. We now know of "hot spots" in bacterial chromosomes wherein mutations occur with an inexplicably high frequency, and other genes (some are called "mutators") that radically affect the likelihood of mutations occurring throughout the genome. Moreover, the difficulty involved in the discovery that a mutation has, in fact, occurred creates a further gap in our certainty. "Seeing" a mutation may require that a bacterium or phage be allowed to divide hundreds of times before any defective product is measurable. By this time, other mutations almost certainly will have supervened, including the possibility of a "back" mutation in which the gene reverts to "normal."

In all this, what emerges is the existence of as yet unseen mercurial forces that work at the level of the gene. The genetic code itself, though well within the clockwork model, operates with a kind of uncertainty. Often only a "wobble" hypothesis, which describes the probabilities of specific base sequences being "read" by a special translating molecule, adequately explains how the code actually works to reduce uncertainty, while redundancy and repetitive sequences of the simple four-letter code further compound the problem of certainty.

If the current microbiological picture of genetic "mechanisms" is paradoxical, the macro view is just as uncertain. Problems of perspective (macro versus micro) and interactional effects are most acute when we come to the study of human genetics. For one thing, analysis of the genetic spectrum of higher functions necessarily leaves out some important and often crucial features of the organism.

Abstraction at this level necessarily results in the loss of information. According to philosophers Alasdair MacIntyre and Samuel Gorovitz, entities such as human beings cannot be understood solely as the sum total of the physical and chemical forces that operate on them. They go on to point out that we cannot expect to be able to move from a theoretical knowledge of the relevant laws of biology to a prediction of behavior.

Some contemporary behavioral geneticists take a more sanguine view of our chances of solving these knowledge problems. Benson Ginsburg of the University of Connecticut, for example, believes that understanding behavior in terms of its biological substratum interacting with experience is "as easy as stopping smoking. Both have been done many times." Beyond this and other tributes to the human spirit, however, the problem seems less one of perseverence or ignorance and more one of our capacity to acquire information and then integrate it.

Physicist and philosopher Michael Polanyi points out the double bind we are in when we wish to know something about a "coherent entity" like a person. We rely on our tacit understanding and awareness of the particular properties that make up the individual in question, but if we switch our attention to these particulars, the synthetic *Gestalt* by which they interact to give us our sense impressions is canceled, and, in Polanyi's words, "We lose sight of the entity to which we had attended." One of the critical risks of genetic research is mistakenly seeing the whole as a direct manifestation of its parts.

There is more to Polanyi's objection than a mere restatement of the idea that the complex properties of an organism cannot be reassembled by adding up all its constituents. Polanyi implies that there are new features that emerge at higher levels of organization and cannot be predicted by the properties of its components. For instance, knowing that a young boy has inherited the gene for classic hemophilia from his mother (it is already possible to make this test prenatally) does not tell us that the boy will be a severe "bleeder." All one knows is that as a result of generating a defective molecule (a variant of factor VIII), there is a much reduced *range of reaction* of the blood-coagulation system. But whether the affected child will be among the 5 percent of hemophiliacs requiring constant prophylactic treatment or among those needing it only periodically or occasionally, cannot be ascertained on the basis of the genes alone.

And, as we have seen, increasing numbers of markers in the blood, like the HLA B27 antigen, can be linked to future devastating disease processes. Unfortunately, such markers have been mistaken as causal agents in public dramatizations of their significance. Such a depiction carries the sense that something like B27 actually causes ankylosing spondylitis and the complex disease processes associated with it. While it has rarely been found that what initially appears as an antigen is, in fact, a causative agent (this was true with a blood protein called *Australia antigen*, which later was found to be the viral agent of hepatitis), it is extremely unlikely that all the HLA markers are the immediate causes of disease. Instead, we have to reorient our thinking about markers in a more holistic sense.

Contemporary scientists recognize that statistical associations between markers (or the genes that produce them, for that matter) and later events or disease processes may be *all they will ever be able to establish*. Such a view is dramatically unlike the past thinking about genes; not so long ago, a statistical correlation was almost always seen as just a first step toward uncovering the actual first causes of a phenomenon. While it was true that Mendel's ratios, in fact, represented the assortment of individual particles or genes, whether or not the gross statistical correlations between clusters and groups of genes we are uncovering today will ever be reduced to an additive picture of simple causation is highly uncertain. If the environment cannot be made perfectly random for all the persons or families whose attributes are being measured, classicial population geneticists like N. E. Morton have emphasized that any seemingly additive effects of individual genes, their interaction or the effects of assortative mating must be neglected; they are simply unmeasurable under such circumstances.

Understanding how an organism copes with external interferences with its well-being opens a new avenue for understanding how human invention further confounds gene action. An embryo cannot have the intention of repairing or buffering the deleterious effects of a gene; similarly, when a nonhuman, noncognitive organism (i.e., one incapable of using learned behavior in its adaptations) copes with a new environmental assault, it is limited by its gene-directed adaptive capacities. Occasionally, the animal will have been prepared for the encounter by prior exposure. A memory of a noxious event can lead to a conditioned response, or exposure to bacteria and viruses can create a preexisting immunity. But ordinarily the organism must deal with each fresh assault by pushing out the invader and repairing the damage anew.

Humans, however, select their responses to insult and, in some measure, even their environments. As Swiss psychologist Jean Piaget has convincingly shown, intellectual development occurs between poles of adjustment to the environment and incorporation of the environment into the person's existing mental structure. The ongoing structural formation of intellectual functions, according to Piaget, is "due to the organism as much as it is due to the pressures from the stimuli, the environment." For example, newborn babies are apparently programmed by their genes to react to the human face on a primitive stimulus-response basis. This inborn reflex may explain why many mothers insist their babies recognize them when, from a neurological point of view, the baby's capacity to do so is very doubtful indeed. Nevertheless, the mother is likely to increase her level of interaction with the child in response to the baby's supposed recognition of her. Thus, the neonate releases maternal behavior that will feed its own social and intellectual development and, so, actively participates in generating the maternal-infant bond.

Besides enhancing development, the human capacity to manipulate, to select, and *to know* can minimize the consequences of insult. Immunization can prepare humans against bacterial infection, and appropriately monitored immunization of a whole population can *erase* the genetic differences in susceptibility to invasion that would otherwise exist. Dietary treatment for diabetes, similarly, can override subtle differences between individuals with genetic predilection toward glucose intolerance. And if interventions such as these can effectively swamp the biomedical effects of genes, why shouldn't education effect similar changes in the genes that presumptively underlie the abilities for learned behavior?

In fact, even our currently rudimentary knowledge of the ways in which higher structures depend on environmental stimulation has led to remarkable advances in the treatment of developmental deficits. Restructuring of the primitive relationships between motor and cortical pathways in deficient adults through forced crawling has been tried in cases where reading skills are impaired. Reinforcement of visual pathways has also been attempted, to overcome genetic or behavioral deficits. Visually rich environments and specific visual stimulation have been associated with improvement of retarded individuals with Down syndrome.

Human adaptability diminishes our certainty in predicting the final effects of genes on biologically complex organisms and, for that matter, delimits our capacity to say which genes will be present in future generations. The muscles of an Olympic weight lifter, the neuronal circuitry of a Rubinstein or a Heifetz, or the oxygen-carrying capacity of high-altitude Peruvian runners, are all extrapolations of genetic possibilities to be sure. But because people act to select the stimuli that impinge on them (no matter in how cybernetic or mechanical a way), the outcome of development defies prediction. Some genes may be left at the starting gate because the preconditions for their activation never occur. Others may be turned on only briefly, their products left waiting for the necessary inputs from the environment to become functional.

Indeterminacy, therefore, should become a basic principle of modern genetics; and in a more circumscribed sense, it can be seen as a basic characteristic of the gene as well. This is true not only in the infinitely complex rules governing the process of genetic assortment in reproduction, but also in the effects of single genes. Many, if not most, genes simply delineate a "range of action" for an organism. With the exception of the relatively rare single-gene–determined conditions in human disease processes, human ability and disability are most unlikely to fall into a simple one-gene-one-effect relationship. Our understanding is greatest where physiology alone is involved; but when psychology enters the picture, possibilities multiply like flies from the Nile. For the really important questions facing us about human nature, fortune and society, genetic explanations are not likely to enhance our feeble capacity for biological and psychological prediction. Rather, as physical scientist John Sutherland has put it, "The introduction of genetic factors as determinants inescapably throws human behavior out of the deterministic domain and into the stochastic or indeterminate."

11

Humanizing the Genetic Enterprise

> What the scientist must now admit is that in many problems of great consequence to people, reality may not be accessible, in practice, through entirely manipulative and analytical methods.
> —ROBERT WALGATE in the *New Scientist* (1975)

Genetic knowledge, when it applies to ourselves, can touch us deeply; and when it applies to others, it can become a too-ready index for social exclusion. In a sense, this new knowledge has caught us unaware. We have no system of common thought that reflects the often paradoxical nature of determinism and free will that seems to characterize gene-directed development. Partly because of the very complexity of the problem, partly too because of the bad associations conjured up by the suggestion that research must be limited and guided by ethical values, little systematic thought has been given to the political and moral significance of the new genetics. Yet, increasingly, the drift of such research raises moral dilemmas in public policy.

Recent research has suggested, for example, that the *Pi*

gene series has important effects on the development of respiratory dysfunction. There are currently twenty-three different alleles, and hence many possible combinations, at this one locus. The most ominous is the ZZ combination. Some 10 to 15 percent of people with this homozygous combination of genes die of childhood cirrhosis of the liver, while virtually all the survivors succumb to pulmonary emphysema in adulthood. The ZZ combination itself is relatively rare, occurring in something like one in every 4,000 to 6,000 individuals at birth, but the heterozygous condition (where one Z gene may appear in combination with a more benign F, S or M variety) seems to occur in one of every 32 to 40 infants born in the United States. The actual health consequences of carrying this Z gene in conjunction with its more benign counterparts are unknown, but some researchers suspect that carriers may be at increased risk for respiratory diseases when compared to the population as a whole. An analogous and equally ambiguous situation exists for asbestos workers, in whom the risk of asbestosis is linked (albeit uncertainly) with the B27 marker. If proved, these two examples alone would identify hundreds of thousands of workers in the labor force as being at appreciable risk for disease.

On a policy level, this genetic knowledge would seem to mandate a program of rapid diagnosis of the genetic condition, combined with preventive treatment, in the form of reduced exposure to known environmental stimulants of respiratory problems (e.g., dust, fumes, smoke and other particulate air pollutants). Note that these presumably predisposing genotypes are just that—*presumed* associations, as yet not firmly established as real links to disease. But what direction should such a program pursue?

Certainly an indeterminate portion of such presumptive gene-related illness may be potentially preventable by minimizing exposure to potentially harmful air-borne pollutants. But do industrial employers have an obligation to lower their ambient-air standards to accommodate the relatively rare ZZ individuals? Or, since no more than one in every 3,600 (the survivors with the ZZ alleles who reach adulthood) will actually have the genetic makeup that is known to put them at extreme risk, can they legitimately maintain that the request to lower "safe" levels to accommodate these few individuals is burdensome and unfair? The countervailing argument, of course, is that there is some fundamental obligation to assure that the workplace is acceptably safe for all who might wish to be employed. Handicapped workers are currently pressing (and winning) this right in many public institutions in the United States, and the H.E.W. has recognized the legitimacy of the claim of the physically impaired that the workplace adapt to them, rather than the other way around. This argument could be extended to those with genetic handicaps. But resolving this dilemma requires care. A federal appeals court has already undermined any such trend by ruling that a worker with hemoglobin S-C disease (a variant of sickle-cell disease) who was dismissed by a chemical company because of his illness, could *not* claim protection under the Civil Rights Act of 1964, since this Act did not protect persons with "ethnic diseases."

For our purposes, there are three general approaches to be considered. One may be called the *fatalistic*. Here, we simply recognize the existence of a particular vulnerability—say the genetic predisposition to lung disease—as a quirk of fate to which any of us might fall victim, and for which society cannot be held responsible. Another approach to policy formation can be called *individualistic*. Here, the individual is assisted by social institutions, first to a better understanding of the problem, and then to optional means of dealing with it. Finally, there is a third approach of *social-welfare activism*. According to this view, those principles of justice and equity that apply elsewhere in society apply to the genetically afflicted individual. For instance, it might be agreed that if conditions are not safe for a specific genetically at-risk person in society, they are (with exceptions) probably inimical to the rest of us too, and we therefore have a collective obligation to correct them.

The Fatalist Approach

Each of these approaches is persuasive in a different way, and each has attracted powerful advocates. Until recently, the fatalistic view has been the most general, and it is largely through the agency of Western science and technology that new options have developed. Paradoxically, fatalism is frequently chosen in Western societies for philosophic or economic reasons. Some workers in genetics believe that the recent explosion of genetic knowledge and genetic-disease hypotheses will not change our habitual approaches to health problems very much—that fatalism has a better chance of revising the new genetics than vice versa. Barton Childs of Johns Hopkins, for example, believes it is futile to hope that society can make the world safe for all the varieties of genetically vulnerable individuals we will be discovering. According to Childs, while we will increasingly find that pollutants selectively injure those with gene-determined susceptibilities, the polluters cannot be saddled with the onus of protection. We cannot, he holds, expect the public to respect egalitarian doctrines in the face of a widespread knowledge of genetic differences, since public attention to the commonweal is "not notably evident in modern times."

Such moral reasoning suggests that the new genetic knowledge can indeed become dangerous and be used as a powerful argument for the fatalistic view. Obesity is an example of an exceedingly complicated health problem that seems to lend itself, prima facie, to genetic fatalism. Many people have observed that it seems to run in families—that is, one frequently sees weight problems not just in a single family member, but as a characteristic of most or all family members. Modern genetics has contributed a possible explanation of this phenomenon, according to which obesity is the result of genetically determined differences in either the size, or the number of somatic fat cells. Genes establish a kind of "set level" for lipids in the cells, according to one such hypothesis. Thus, when a "genetically obese" person diets, the cells may temporarily burn up lipids at an increased rate; but, as the lipid concentrations fall below the set level, the brain alters feeding behavior so that the level is restored. In this model, then, an overweight individual is fated to maintain his or her phenotype.

There are, of course, problems with the fatalistic point of view. For one thing, genetic obesity—like other "genetically based" human attributes—is so far only a hypothesis; it has not been convincingly proved or dismissed; and it will undoubtedly be very difficult to dispose of it one way or the other. Obesity, like IQ, does appear to run in families. Parents with gradations of obesity, from very lean to obese, have children who closely mimic their parents' phenotype just as one would expect from a genetic model of causation; where both parents are lean, the children are characteristically thin, and the converse.

Only recently has it been shown that these very same relationships hold for the relative obesity of *adopted* children and their nonbiological parents, thereby casting doubt on the validity of the idea that what was being observed was in fact a genetic phenomenon. Families are partly defined by their sharing of a common environment and a common set of behavioral or psychological styles. Thus, fat pet owners tend to have fat pets; yet, we cannot assume in this case that genes have much to do with this "family resemblance."

In many cases, acceptance of the fatalistic genetic view is all that is needed to make it come true. Telling obese parents that their overweight child is biologically a "fat person" can condemn that child to failure in future weight-control efforts. The very power of genetic attribution can paralyze attempts to treat the problem. In fact, the impotence of a particular treatment style may arise because patients come to believe that they are battling the biologically inevitable, while researchers are discouraged from developing the maximally powerful treatment procedures. In various behavior-related disorders of human equilibrium, the genetic hypothesis supports and strengthens the individual's sense of inferiority or helplessness, and itself becomes part of the problem.

Individualistic Approaches

The individualistic approach recognizes genetic differences as the basis for public policy. Unlike the fatalistic approach, which tends to adopt an overly simplistic, one-gene-one-effect view of genetic determination, the individualistic view is on target so far as one relies on an understanding of the degree to which a particular genetic makeup changes a single person's prospects for disease. Alterations of life-style or environmental conditions to minimize the likelihood of disease are the hallmarks of this model.

In terms of both behavioral potential and disease response, there is every reason to expect critical individual variability. The evidence for genetic individuality discussed in Chapter 2 points to a vast genetic reservoir of variation for the species as a whole. Those of us who carry deleterious genes, or for whom the 6.7 percent heterozygosity embraces genes associated with disease, will likely fall at one end of the spectrum of susceptibility to disease. Equally likely is the existence of a significant portion of the population with a better than average outlook for well-being. In time, individually defined genetic profiles will be the rule, not the exception. Just as we now use blood "serotypes" to range the 16 or so major blood group substances, we will more than likely use the HLA haplotypes, hemoglobinopathies, hyperlipidemias and a host of metabolic disease markers to describe individual predilection for disease. But is such a future a desired one? you might ask.

In certain political contexts, this principle of individualism can serve as the best guide to the formation of public policy. From this perspective, the primary obligation of society is to educate individuals to their genetic susceptibilities and to the options available for modulating these susceptibilities effectively. The premise of such education is that different genotypes represent different adaptations to the environment, and hence to health more generally. We would find significant numbers of people (if we are to accept the hypothesis offered by Dr. Irwin Bross, a radiobiologist at the Roswell Park Memorial Institute in Buffalo, New York) with markedly different susceptibilities to radiation damage. Other evidence supports the existence of persons with alleles which predispose to respiratory illness.

One solution to the problem of such genetically based respiratory dysfunction, for example, would be to identify, and familiarize affected individuals with, the substances likely to trigger major problems. This is the approach currently employed in Sweden, where virtually every newborn is screened for the presence of the enzyme marker of the Z allele presumed to be associated with heightened risk of respiratory illness (the MZ, SZ and FZ phenotypes are intermediate in risk to ZZ). Even though this newborn-screening approach has been largely rejected in this country for its ambiguous predictive value, the people of Sweden seem to take it seriously. Recommendations for appropriate occupations are made early and it is up to the individual parents to act on them.

Given the fact that Swedish air-pollution laws are uniform and conservative (in the sense of high), and given the fact too that occupational choice in Sweden is relatively unaffected by such variables as race and social standing, this reliance on individual control seems a reasonable one. But in other societies, notably our own, discretionary regulation is more problematic. The basic question is the individual's capacity to respond to the educational approach common to the health promotional activities so widely encouraged by health planners—and so poorly subscribed to by lay citizens. It is one thing to advise a man with the gene for an enzyme known as glucose-6-phosphate dehydrogenase (G6PD) not to eat fava beans because they may poison him; it is quite another to ask someone with one Z gene to avoid fumes that are widely dispersed in the air we breathe, or that may occur in sublethal concentrations in particular job settings or sections of a city or town. Where beans are a necessary staple of life for the at-risk black population with the highest levels of G6PD, and employment in the cement factory the only job in town, it is somewhat beside the point to issue discretionary warnings. In these cases, placing the burden of regulation on the individual seems more in the service of particular economic or political interests than in the service of public health. Another possibility then is a public policy which shifts the burden of proof for creating conditions tailored to individual genetic differences from the citizen to the government.

The Social-welfare Activism Approach

This brings us to a consideration of the third major approach. Establishment of socially mediated environmental standards would be an example of this approach. John Rawls has defined a principle that can be used to undergird the basic ethical assumption of activism, namely, what he terms in his book *A Principle of Justice* the "difference principle." This view holds that where conditions are found to be unequal, or natural differences contribute to inequalities, they should be rectified in favor of the least-well-off person. At the hub of this principle is the premise that we should operate behind a "veil of ignorance," another way of saying "there but for the grace of God go I." In a genetic sense, this axiom is particularly apt, since the vagaries of sexual recombination in mating could well endow our own offspring with adverse genetic properties.

Extending this difference principle to the environment, it may be said that redressing regional differences in air-pollution standards should be made with an eye toward the population at greatest risk. While this does not necessarily have to be the *rare individual* who is homozygous for a susceptibility gene, genetic makeup nonetheless provides an index against which to judge "safe" standards. This principle was universalized by Richard Wilson, of the Energy and Earth Policy Center at Harvard when he stated that "What we can reasonably demand is that no worker by virtue of his job, have a risk appreciably greater than other workers."*

* Shifting the emphasis of the sentence from "job" to "worker" conveys the sense intended here: i.e., "no worker whose genetic makeup permitted him to get the job in the first place should, *by virtue of that makeup,* have a risk appreciably greater than other workers."

The Chinese appear to have incorporated at least a related principle to their policy making during the heyday of the Mao-inspired revolution: if conditions cannot be made safe or healthful toward the well-being of the least-well-off person, they should not be tolerated by the rest. Sinologist Joseph Needham has confirmed this value orientation. According to Needham, it would be unthinkable in Maoist China to consider a person a "walking flask of amino acids" genetically isolated from his peers.

Individuality takes on a different cast in the political system developed after the Revolution—each person relates his own existence to that of his co-workers. In pre-1978 China, economist John Gurley has observed, development was not valued unless there was a sense of everyone rising together. No person was to be assigned a higher—or lower—intrinsic worth merely because of idiosyncratic differences, be they genetic or social or physical. In principle, no one was to be left behind if the ideals of the Revolution were to succeed.

In practice, the situation undoubtedly drifts away from this ideal, but the actual reports of persons who have gone to China in the early 1970s to examine how this principle has been applied are remarkable for their consistency. An anecdote told by educational specialist Urie Bronfenbrenner, of Cornell University, nicely illustrates the jarring difference between this view and others to which we are accustomed. Bronfenbrenner once pointed out to his Chinese hosts that he had seen extraordinary effort put into the study of rice breeding at a local agricultural field station, and equally great care taken in the matching of irrigation and fertilization schedules to the requirements of different rice strains.

"Why, then," he asked, "if you put this much attention into the ideal conditions for the flourishing of your plant varieties, don't you do the same thing for your children?"

Bronfenbrenner's guide replied, simply, "All plants are different from each other, but it is important to believe that all children are the same."

This dialogue, which at first hearing seems merely Confucian or antiscientific, actually addresses some of the most fundamental issues in genetics and policy planning. Bronfenbrenner begins from a point of view that one finds consistently in lay misunderstandings of the new genetics—that is, from the point of view of a Mendel trying to understand the meaning of human differences and trying to develop means of coping with them. His syllogism might well begin (implicitly), "All men are like peas." His Chinese host gently corrects him, perhaps because he starts from a Marxist perspective of radical environmentalism, "If you treat men as equals, then you will produce a society of equals." Or, to put it in terms familiar to Johannsen and students of phenotypes, "If you take genetically dissimilar pea plants and place them in a field with an optimum set of trace nutrients, they can still all grow to the same average height."

This view is countered by such Western critics as developmental psychologist Gardner Lindzey, who criticizes the Chinese approach as being fundamentally antiprogressive. Equality, according to Lindzey, may be wished for and even worked at, but "it is dangerous in the extreme to assume in advance that [biological equality] is so." According to this view, assuming equality in the face of manifest, socially induced inequalities—*as well as* genetic ones—may be one of the best ways of perpetuating inequality.

What, then, should we do with the overwhelming reality of human differences? First, the argument may be made that knowing *in general* how much genes condition the prospects for improvement or maintaining the status quo can be critical. For instance, some researchers have concluded that governments can better design their programs if they know which variations in the environment can make a difference for a given group—and which cannot. When we get down to the individual, genetic realities may indeed be the overarching ones. No one, for instance, would deny that children with Down syndrome or dominantly inherited severe mental retardation have special educational needs and ought to be given special services. In a moral sense, the biological perspective has some relevance: Respect for individual differences is at root a derivative of a genetic perspective. But as long as that perspective means that labels will be applied, the respect will be far off.

Public-policy makers are increasingly finding that the more label-free is the environment in which you place developmentally handicapped individuals, the more dramatically you can "normalize" their social or educational experiences. Socialization often can occur *only* in normal environments—and here genetic differences fall to the Chinese ideal.

Second, we need to recognize that each of the three approaches to genetic policy planning discussed here can be useful in specific situations. Even the most dubious approach—the fatalistic—has its place. But currently, the great preponderance of information generated by programs like multiphasic genetic screening is therapeutically useless. We simply don't know what to do with this knowledge in terms of helping the individuals who are most affected. Given the propensity to misunderstand genetic "determinism"; given the profound psychological impact of such misunderstandings; and given the tendency of genetic explanations to effect self-fulfilling prophecies, the continued emphasis on knowledge as an end in itself is a doubtful proposition.

There may be situations wherein simple good sense—or

prudence—says it is best *not* to know. In sickle-cell–anemia screening, for instance, it has occurred that a child's genetic makeup (based on the presence or absence of sickle-cell genes) is found to be inconsistent with the supposed paternity of the child. Experience has shown circumstances where such unasked-for information can do more harm than good. Similarly, deciding not to go after some genetic information, especially when that "information" is ambiguous, may be justified by social as well as individual reasons. There may be times when moral or political considerations lead us to consider retaining a veil of ignorance over the suspected genetic contribution to group differences in a behavior or attribute in which the measuring tests are suspect. And I am thinking here about the so-called "measured" genetic basis for IQ differentials between blacks and whites. We say "no" to this form of inquiry, not because such knowledge is in itself evil, but because in many cases it is simply misleading and self-defeating.

Given the proved existence of systematic, *non*genetic differences between the two groups that could account for the different heritabilities of IQ scores, pursuit of further genetic tests in lieu of environmental testing or remediation is scientifically invidious. Certainly, in the first instance, it is reasonable to ask about the moral consequences of uncovering the "information" sought, as well as the legitimacy of asking for it in the first place.

The consequences of verifying (or rejecting) a genetic hypothesis for a disease process are also shot through with social implications that are rarely considered. Admittedly, being able to *reject* a genetic hypothesis may have tremendous social value. But coming up with the genetic idea and insisting that it be tested first may delay the day when meaningful environmental approaches to a disease are tried. Genetic hypotheses often emerge from only dimly perceived cultural and political orientations in the mind of the investigator, as we saw at the very beginning of this book in the case of pellagra and undernourished Southern blacks and whites.

Genetic models may undercut the incentive for any medical intervention at all. If low hemoglobin levels in the black population are "hereditary," as two researchers now propose, then the urgency of finding a solution for chronic anemia among the rural and urban black population is greatly dampened. And these same researchers ask their colleagues to consider the possibility that low birth weight among blacks is hereditary. Given that low birth weight accounts for most of the variance in subsequent morbidity and mortality, relegating black newborns to a hereditary norm could very well prematurely assign them to a kind of medical twilight zone, wherein interventions to improve birth-weight scores beyond the black "ideal" are abandoned.

Finally, the realities of genetic variation and distribution in the population suggest a resolution of the problem. The range of human potential, disability and vulnerability for a wide variety of traits and attributes is described by the bell-shaped curve of normal distribution.

Qualities as diverse as blood pressure and height follow this configuration. A variant of this curve, developed by the French mathematician Poisson, showed that even a cavalryman's chances of being kicked by a mule follow a "random" distribution of probabilities for rare events.

At each end of any normal distribution—that is, at the tails of the bell-shaped curve—are areas in which we may wish to focus attention on genetic factors. And it is here that confounding events are least likely to skew our ability to recognize genetic differences. But what of the middle of the curve, where the great bulk of people are to be found? How far are we obliged to dissect the bell in search of genetic contributions to minor differences? Are we not, like the Chinese, better off behaving as though all persons are equal in potential, so far as most human qualities are concerned?

Take for instance the case of the cancer-prone carriers of the genes for the DNA repair disorders which we discussed in Chapter 3. In time, it might prove valuable to be able to detect these persons outside of the rare family setting where we now find them. But would we *need* to find them to make our cancer policies? Currently, we must be including all of these persons—with their dramatic predilection toward cancer—in our statistics. Thus we are unknowingly factoring in these high risk groups with both morbidity and mortality figures *and* our cancer policies. If anything, the Rawlsian approach would have us adjust our standards downward to protect the vulnerable one in 100 or so with the susceptibility gene, a policy I endorse.

Some would say that we could effect relatively large societal savings by testing for—and excluding—such carriers in industrial settings, or encouraging their resettlement should they be in high pollution neighborhoods or towns. And don't we owe these persons the right to know that they are different—and more susceptible to the risks of living generally, and to sunlight particularly?

First, no one has proposed that the mechanism of carcinogenesis is qualitatively different in carriers—and the rest of us (assuming we know *we* are not carriers). Should we reduce the strictures on environmental exposure to carcinogens for the population at large, we would be increasing all of the non-carriers' marginal risk of cancer too. If we set pollution control standards based on some new population norm, we would simply be putting a *new* group on the tails of the bell-shaped curve, and we would have to start the whole exercise all over again.

Second, if we could detect such carriers, wouldn't we

rather use them as the *standard* for cancer risk? Or do we allow their unique genotype for cancer susceptibility to put them beyond the requirement for equal protection? The veil of ignorance is particularly apt here. In time, we would have to face the prospect of spontaneous mutation or in-marriage bringing the same susceptibility genes into our own lineage.

Until such time as we have the aggregate population partitioned into such susceptibility categories, we will unknowingly perpetuate just the kind of policy which the Chinese advocated by behaving as if genetic differences don't exist for the purpose of carcinogen policy. Were we to decide in the face of the information on susceptibility to keep the standards the same, while perhaps providing extra precautions for each individual found to be at risk, we would be deciding against the Chinese view to overlook genetic differences. And, you might argue, we would have made the decision consciously, informed by the force of arguments of social justice to act to protect the vulnerable population in our midst. But are there circumstances where the Chinese model is appropriate?

One consequence of such radical egalitarianism is that it neglects the differences that deserve redress or compensation, and it is here that I favor Rawls's viewpoint.

On a policy level, we have already applied this belief in such areas as taxation and disability training; but we are far from doing so in terms of ensuring equal educational and health-care opportunity. And here is the rub: Should we not acknowledge some obligation to consider the disrupting effects of circumstance, the injustices imposed by historical forces of oppression, and even the natural inequities attendant on birth? We now know, for instance, that virtually every health indicator—from low hemoglobin level to raised blood lead levels, indeed every item measured on standard laboratory tests of nutritional status*—

* Specifically hemoglobin, riboflavin, vitamin A and vitamin C.

varies systematically according to income, with low-income persons having consistently poor outcomes. If we give weight to these factors, Rawls's difference principle takes on increased moral consequence. Extending this principle from health and educational policy into environmental planning, it may be said that uniform air-pollution standards (for example) ought to be established across states and regions; and that these standards should be written with an eye toward the population at greatest risk—*both* historically and as determined by their station in life—and their genes. In the face of ignorance of these factors, say for resistance to radiation damage, aren't we better off by putting the standard at zero—as it might be for the homozygote deficient for DNA repair enzymes? Similarly, where, in the tails of the bell curve, *known* genetic factors may be found to predominate in a way that overrides our present-day ability to control hazards, then the individualistic approach of specifically tailored environments appears justified. And these approaches would be made for those at both extremes—the *very* gifted, for whom everyday environments are an oppression—and the *very* severely impaired, for whom the richness of the environment surrounding them may be the only hope for a modicum of decency. Down syndrome children, for instance, thrive (relatively speaking) when their mothers have special training to enrich their environments, as compared with similar infants left at home without this special attention. (See Chapter 6 for a fuller discussion.)

For the future, we must consider policy planning in biomedical research, specifically in areas in which competing hypotheses of causation are equally uncertain. In my view, as long as we are dealing with problems that affect one generation, with groups involving large numbers of people, our first approach to unraveling the complex mix of factors at work in a given disease process ought to be (as a first approximation) an investigation of environmental cause. Not that genes might not be at work beneath the surface, but merely because an environmental hypothesis provides us with greater freedom in our actions to institute therapeutic options than does a purely genetic one. Eventually, the most productive analysis would be directed at the *interface* between environments and genes.

This view is based in part on what we have learned about the characteristics of genetic contributions to major disease entities like cancer, hypertension and heart disease: while an appreciable portion of each of these entities has a distinct genetic underpinning—and a few specific syndromes and entities have a direct, single-gene-cause—the majority in each classification falls into the broad center of the bell-shaped curve where complex interactions, probably of multiple alleles with multiple external factors, act to shape a normal distribution.

In the end it may be impossible to sort out the contributions of genes and environment to specific traits—not because of problems of measurement, but because of the confounding effects on genetic programs by random environmental events. Thus, a residue of uncertainty seems to exist for each of us with regard to our genetic heritage; and this is more than a by-product of the finiteness of our information. This residue needs to be incorporated into public policy, to take account of the differences between the individuals served by educational or health programs. This indeterminacy almost certainly underlies the expression of our most complex traits. And it is this indeterminacy that creates the space for enlightened policy making.

12

Conclusion

> The complexity of gene action at the molecular level . . . require(s) that we give careful attention to . . . the finality with which genetic information determines traits of the most urgent human interest.
> —J.C. Loehlin, G. Lindzey and J.N. Spuhler, *Race Differences in Intelligence* (1975)

In the end, we may find that we learn much or sorry little about our own special needs by plumbing our genetic makeup. But unless we have solid theoretical backing for each research enterprise—an understanding of the limits of what the genetics involved can tell us—we will come off flat-footed in our science. Worse still, unless we make some assessment of the moral and political costs of research, we may find that the cost of the exploration will outrun the possible benefits.

It may turn out that the notion that genetics can tell us something meaningful about complex human behavior is wrong to begin with. If this is so, as the preliminary analysis presented in this book suggests, then the legitimacy of further inquiry into the presumptive genetic basis of such complex human traits as altruism or criminal deviance is in question. Accordingly, we should be prepared to assign priority elsewhere.

We are under a veil of ignorance, to be sure. But we can say this: at this point in our evolving ability to understand the forces that shape human nature, it is premature to look to genetics as the source of the "truth" about such matters as the expression of human aggression and intelligence.

We must wonder, if absolute truth does not exist in genetics, will some still find it necessary to invent it? The historical abuse of genetics suggests that the answer, in some quarters at least, will inevitably be yes. Powerful social expectations shaped the uses of genetic knowledge following the rediscovery of Mendel's work in 1900. As we have seen, susceptibility to tuberculosis, mania, pellagra, criminality, prostitution, imbecility and alcoholism were all at one time or another unequivocally attributed to individual genes. It was years before researchers realized that social problems, not genetics, were at the root cause of such ills. Genes undoubtedly play a contributory role to disease susceptibility, but in turn-of-the-century America, poverty was surely the root cause. And for many of the traits and conditions which are today assigned a genetic legacy, poverty is still the ultimate arbiter, be they IQ scores, infectious disease susceptibility, behavioral deviance, or congenital malformations.

Nor should we assume today that the popularity of genetic models in turn-of-the-century America was confined to socially conservative factions. Indeed, it was the progressive movement of the 1920s that turned to eugenic programs as an expedient answer to the inequities of class. Fortunately, the selective sterilization programs of this group were doomed by the complexities of genetics itself and because these policies threatened fundamental institutions like the Catholic Church. But this does not mean that the rationale for sterilization policies based on genetic knowledge does not still exist. Sixteen states still have on their books legislation that permits selective sterilization of the mentally incompetent. Most do not use it, but the shadow of abuse remains.

Today, we presume that genetic research is free from the biases and most of the poorer methodologies that plagued earlier work. We think we recognize the serious mistakes made by researchers when they omitted social, dietary or other key environmental factors from their analyses of the roots of social problems. But do we recognize the contemporary forces that confound genetic research?

Political usurpation of genetics will not be stopped by reciting a litany of past abuses, or by careful dissection of spurious research designs or methodologies. Abuse will recur whenever people in positions of power can freely bend scientific half-truths to political ends or subconsciously manipulate genetic data to shore up particular predilections or ideological biases. And scientists who wish to thwart such abuse must exert a new measure of control over their work and take responsibility for its interpretation. Those who work under totalitarian regimes rarely have the opportunity to do so. But those who practice science in a free society have greater prospect of control. The question is how.

It is likely that in time the historical record will again show that where genetics has been misused, it will have been because scientists themselves have failed to show forcefully where interpretation of their data should stop or to voluntarily limit their work when it posed a clear threat of abuse. A remarkable and notable contemporary exception to this pattern was the initial willingness of scientists who work with recombinant DNA molecules to restrict their research while the minimal conditions for safety were worked out. But, as with the nuclear physicists, it has proved extremely difficult for these or any other scientists to accept that there are moral reasons for not going after some knowledge. Recognizing limits to scientific experimentation entails at least acknowledging the conceptual boundaries of scientific explanation, if not the moral restrictions on conducting scientific inquiry where the risks of harm or other deleterious consequences can be foreseen. As Bernard Shaw put it in *The Doctor's Dilemma*, "No man is allowed to put his mother in the stove because he desires to know how long an adult woman will survive the temperature of 500 degrees Fahrenheit, no matter how important or interesting that particular addition to the store of human knowledge may be."

We have seen abuses of the research imperative in our own time in the initially unconsenting use of newborns in Baltimore and Boston for chromosome analyses in the hopes of finding a relationship between behavioral deviance and the presence of an extra Y chromosome. We have seen compulsory mass-research screening programs for such genetic markers as sickle-cell hemoglobin initiated without clear-cut benefits to the participants. We have seen genetic "theories" of intelligence take center stage while valid environmental hypotheses—like the contribution of lead to IQ deficits and behavioral disturbances —have had to wait in the wings for their testing. And we have seen genetic tests used arbitrarily to refuse health-insurance coverage or employment.

How can such abuses be avoided in the future? "Criminal" identification through chromosome analyses, preferential education for the genetically "better endowed," and diversion of badly needed medical resources into spurious screening programs represent our present excesses. But these and future abuses of genetic research will be likely to recur until the basic premises of genetics are critically reexamined, and the limits to genetic interpretation are acknowledged by geneticists and policy makers alike. Geneticists have always known, or strongly suspected, that the phenomena that they were studying were complex—indeed, that is an axiom of genetic science—but they drew back from making this complexity known to the people—like myself—whose decisions affect the lives of others. It is time for scientists to speak out wherever and whenever they believe their work is oversimplified or misinterpreted to bolster a particular social policy.

In the meantime, social scientists continue to look to genetics for new paradigms, for reasons that are evident enough—in areas like schizophrenia their own models are felt to be inadequate to explain complex phenomena.

Because of its firm grounding in molecular biology, genetic explanations of these phenomena create an illusion of being more essentially true than some of the relationships established by psychology or sociology. As Bernard Davis put it in a recent symposium on the uses of genetic knowledge, "Social policy based on [genetic] reality will be more valuable than social policy that is based on a shaky foundation. . . ." But genetic models of complex psychocultural-biological disorders like schizophrenia have already been abused in the Soviet Union and promise such abuse in this country unless enlightenment is forthcoming. Complex environmental and psycho-social forces operate in even the most "clearly" genetic case of affective disorders.

There is no prima facie reason to accept or reject the premise that genetically based reality—whatever it happens to be—will necessarily provide a sound basis for social policy, any more than adherence to a moral principle such as distributive justice provides a universal resolution to moral dilemmas. Genetics *may* spell out certain inevitabilities of human life, but it provides no direction for dealing with them. Nor does genetic determination necessarily incorporate the effects of human effort (positive or negative) on the direction of human events.

Until and unless some factual basis for attributing significant human variation in socially significant traits is forthcoming, it is premature to speculate on programs based on the premise of genetic control for these higher functions. For many problems we lack statistical approaches or data-collection methods capable of even asking the questions in the right way, much less providing us with definitive answers. And as long as the statistical measures remain so dubious, we are groping in the dark as far as a genetic basis for social policy is concerned. We know from past experience that such unenlightened policies are likely to hurt someone.

Instead of emphasizing the largely hypothetical problem of tailoring environments to meet presumed individual genetic idiosyncrasies, we ought to be using what biological or social knowledge we already possess. One objective that could be set now is the creation of those conditions that make possible the realization of at least *minimal* levels of ability and well-being for the greatest number of persons. Belaboring a genetic model or approach is undesirable if it shifts attention from the fact that we have not yet created an environment that reflects what we *already* know to be necessary for realizing human potential. We know, for instance, that ascorbic acid, or iron deficiency or raised blood lead levels in poor children definitely contribute to the link between socioeconomic condition and intelligence. Reducing the damage done by environmental deprivations like these is realizable with our present understanding.

We also know how to supplement the social conditions (for example, through systematic social enrichment) that we have good reason to believe are prerequisites for the later flowering of human development and potential.

Should we, then, refrain from using genetic knowledge for instituting public policies in education or social reform? Or should we continue to seek that knowledge? The traditional answer, "Let's do both," is unsatisfactory. Genetics has already provided us with critically valuable data about ourselves. And much of it is being applied. In the face of so many "good" uses of genetics—in counseling, amniocentesis, prescriptive screening—it would be tragic to continue to siphon so much energy into such potentially mischievous areas of genetic research as social or behavioral deviance or personality traits. The search for new knowledge in these areas has such very high social cost, and its benefits are so unclear that it may be prudent to give this research a lower priority. To do otherwise is to flirt with the danger of social Darwinism, where scientific half-truths are put to the task of validating political policies. We need to realize that ignorance cannot be dispelled by arrogance.

Nor can we rely solely on the idea that genetics will be used well as long as it is done well. This is part of a view that holds that scientific ideas are innocuous in and of themselves, and that any evil attending them is purely the result of misapplication. It should be clear now that some genetic ideas, such as the notion that criminal behavior is strongly molded by the genes, lend themselves to abuse.

Genetics is too important to be left to the geneticists. We cannot afford to have scientist-lobbyists vying for governmental support or public recognition purely on the strength of their convictions that genes are at the root of all human problems. Genetics may indeed be critical to the fullest understanding of the operation of living things; but, taken in isolation, genetic science is still inadequate to the task of describing the humanistic elements that are most important to us. The solution to these problems may well not be more science, but more careful science; not fewer genetic models, but more complex genetic models; and not unbridled optimism of what genes can do, but rather a tempered, more humble, if not grudging, acceptance of the limits of our ability to know. Genes are not everything.

NOTES

CHAPTER ONE

(pages 11–31)

Claude Levi-Strauss, *The Savage Mind* (University of Chicago Press, Chicago, 1966), provides a broad perspective on the cultural origins of exogamy. The contemporary relevance of this concept is explained in an exceptionally readable article by J. V. Neel, "Lessons from a 'Primitive' People," *Science,* Vol. 170 (1970), 815–22.

For a complete description of the origins and impact of Mendel's ideas, see Uri Lanham, *Origins of Modern Biology* (Columbia University Press, New York, 1968). An excellent review of Galton's relation to modern genetics is found in Barton Childs, "Garrod, Galton and Clinical Medicine," *Yale Journal of Biology and Medicine,* Vol. 46 (1973), 297–313. Galton's biased view of the racial issue was revealed in C. P. Blacker, "Galton's Views on Race," *Eugenic Review,* Vol. 43 (1921), 19–22.

A historical reference for the presumed biological inequality of the races is Charles Loving Brace, *The Races of the Old World* (Scribner, New York, 1863). (For a damning work of scholarship that reveals the limited objectivity of Brace and his contemporaries, see S. J. Gould, "Morton's Ranking of Races by Cranial Capacity," *Science*, Vol. 200 [1978], 503–9.) The steps to the extreme view that races were so dissimilar as not to be capable of interbreeding is reviewed in W. B.

Provine, "Geneticists and the Biology of Race Crossing," *Science,* Vol. 182 (1973), 790–96.

The next phase, in which frank eugenic ideas were put forward, is traced in M. H. Haller, *Eugenics: Hereditarian Attitudes in American Thought* (Rutgers University Press, New Brunswick, N.J., 1963), and in K. M. Ludmerer, *Genetics and American Society* (Johns Hopkins University Press, Baltimore, 1972). Historical references used include H. E. Walker, *Genetics* (Macmillan, New York, 1913); P. Popenoe and R. H. Johnson, *Applied Eugenics* (Macmillan, New York, 1922); A. E. Wiggam, *The Fruit of the Family Tree* (Bobbs-Merrill, Indianapolis, Ind., 1924); and the classic review of that period, H. H. Newman, *Evolution, Genetics and Eugenics* (Greenwood Press, New York, 1932).

The influence of eugenic ideas on social thought is, of course, complex; and it is the source of much controversy. Representative views followed the publication of Richard Hofstadter's controversial *Social Darwinism in American Thought* (Beacon Press, Boston, 1965); see in particular the essays compiled by J. Wilson in *Darwinism and the American Intellectual* (Dorsey Press, Homewood, Ill., 1967). A glaring example of the distortion of scientific thinking by the influence of Darwinian views is traced in Allan Chase's popular "The Great Pellagra Cover-up," *Psychology Today,* February 1975, pp. 83–86.

An extremely reliable and comprehensive treatment of eugenic sterilization is made in Philip Reilly, *Genetics, Law and Social Policy* (Harvard University Press, Cambridge, Mass., 1977). Source materials for this subject include Charles Davenport, *Heredity in Relation to Eugenics* (Henry Holt, New York, 1911); and P. Popenoe and R. H. Johnson, *Applied Eugenics* (Macmillan, New York, 1922).

The history of the use of fallacious genetic arguments for the transmission of susceptibility to tuberculosis is traced in A. R. Smith, *Pathogenesis of Tuberculosis,* Springfield, Ill., 1948. Insurance company policies on hereditary diseases and their confusion with nongeneric disorders can be found in the *Proceedings of the American Life Insurance Medical Directors,* 1889–present.

For general background on genetics, its theory, principles and operations, the reader is referred to William L. Nyhan, *The Heredity Factor: Genes, Chromosomes and You* (Grosset & Dunlap, New York, 1976); a more popularized version can be found in Philip Goldstein, *Genetics Is Easy* (Viking Compass Books, New York, 1967). Classic serious texts are L. L. Cavalli-Sforza and William F. Bodmer, *The Genetics of Human Populations* (W. H. Freeman, San Francisco,

1971); and Curt Stern, *Human Genetics* (W. H. Freeman, San Francisco, 1960).

I relied on two key references for the book—Victor A. McKusick, *Mendelian Inheritance in Man: Catalogs of Autosomal Dominant, Autosomal Recessive and X-Linked Phenotypes,* 5th ed. (Johns Hopkins Press, Baltimore, 1978); and C. J. Epstein and M. S. Golbus, "Prenatal Diagnosis of Genetic Diseases," *American Scientist,* Vol. 65 (1977), 703–11—as well as more technical works for estimates of the incidence of genetic disease in human populations. The state of the art (circa 1977) for identifying genes on human chromosomes is described in V. A. McKusick and F. H. Ruddle, "The Status of the Gene Map of the Human Chromosome," *Science,* Vol. 196 (1977), 390–405.

The interested reader is referred to a number of more accessible articles for a flavor of the uses to which genetics is being put: Maya Pines, "Genetic Profiles Will Put Our Health in Our Own Hands," *Smithsonian,* July 1976, pp. 86–91; Richard Restak, "The Danger of Knowing Too Much," *Psychology Today,* January 1976, pp. 64*ff;* J. Greenfield, "Advances in Genetics That Can Change Your Life," *Today's Health,* December 1973, pp. 20–24; and D. L. Heiserman, "New Hope in the War on Genetic Disease," *Science Digest,* February 1973, pp. 68–72.

A classic text for an explanation of inborn errors of metabolism is Harry Harris, *The Principles of Human Biochemical Genetics,* 2nd Ed. (North-Holland Publishing, Amsterdam, 1975). pp. 187*ff.* A less technical, but nonetheless exacting treatment can be found in Victor A. McKusick, *Human Genetics* (Prentice Hall, Englewood Cliffs, N.J., 1973), pp. 99*ff.*

The phenomenon whereby one of two X chromosomes in the female cell-line is inactivated randomly early in embryonic development and thereby permits the expression of a "normal" X-linked gene to compensate for an abnormal one (as in the example given in Chapter 2 of a woman who carries the gene for hemophilia A) is discussed lucidly in R. W. Erbe, "Principles of Medical Genetics," *New England Journal of Medicine,* Vol. 294 (1976), 381–83. X-inactivation was discovered by Mary Lyon, "Sex Chromatin and Gene Action in the Mammalian X-Chromosome," *American Journal of Human Genetics,* Vol. 14 (1962), 135–7.

The potential problems associated with the expansion of prenatal diagnosis are reviewed in T. M. Powledge and J. Fletcher, "Guidelines for the Ethical, Social and Legal Issues in Prenatal Diagnosis," *New England Journal of Medicine,* Vol. 300 (1979), 168–72.

In discussing occupational susceptibility to disease and its possible genetic basis, I relied on E. S. Calabrese, *Pollutants and High Risk Groups* (John Wiley, New York, 1978) as a general reference. Readers interested in pursuing the topic of nongenetic factors in the genesis of cancer are urged to read R. T. Prehn, "Nongenetic Variability in Susceptibility to Carcinogenesis," *Science*, Vol. 190 (1975), 1095–96. The specific works cited in the text on variability in immune responsiveness are J. Graff, M. A. Lappé and G. D. Snell, "The Influence of the Gonads and the Adrenal Glands on the Immune Response to Skin Grafts," *Transplantation*, Vol. 7 (1969), 105–11; M. A. Lappé et al., "The Importance of Target Size in the Destruction of Skin Grafts with Weak Incompatibility," *Transplantation*, Vol. 7 (1969), 372–77; and M. A. Lappé and J. Schalk, "Necessity of the Spleen for Balanced Secondary Sex Ratios Following Maternal Immunization with Male Antigen," *Transplantation*, Vol. 11 (1971), 491–95.

CHAPTER TWO

(pages 32–52)

For an accessible account of modern genetics and its human dimensions (including the Edgar Winter story), see Adela S. Baer, *The Genetic Perspective* (W. B. Saunders, Philadelphia, 1977). A more technical depiction of the DNA story can be found in François Jacob, *The Logic of Life*, trans. Betty E. Spillman (Random House, New York, 1973), pp. 247*ff*.

The psychological dimension is told in R. J. Stoller, "The Weight of Genetic Knowledge," in *Psychiatry and Genetics: Psychological, Ethical and Legal Considerations*, eds., M. A. Sperber and L. F. Jarvik (Basic Books, New York, 1976). It provides an extremely sensitive analysis of the psychological impact of genetic knowledge. A key article that deals with the psychological dimension of genetics and its meaning is A. F. Korner, "Some Hypotheses Regarding the Significance of Individual Differences at Birth for Later Development," in *The Psychoanalytic Study of the Child* (International Universities Press, New York, 1964), Vol. 19, pp. 58–72. The psychosocial problems generated by genetic counseling and amniocentesis are explored in two articles by John Fletcher, "The Brink: The Parent-Child Bond in the Genetic Revolution," *Theological Studies*, Vol. 33 (1972), 457–85; and "Moral Problems in Genetic Counseling," *Pastoral Psychology*, Vol. 23 (1972) 47–60.

Difficulties in retaining genetic information are documented in C. O. Leonard, G. A. Chase and B. Childs, "Genetic Counseling, a Consumer's View," *New England Journal of Medicine,* Vol. 287 (1972), 433–39. The reported increase in divorces is from an English study, C. O. Carter, H. A. F. Roberts, K. A. Evans and A. R. Buck, "Genetic Clinic: A Follow Up," *Lancet,* Vol. 1 (1971), 281–85; and in A. E. H. Emery, "Social Effects of Genetic Counseling," *British Medical Journal,* March 24, 1973, pp. 724–26.

Specific problems associated with counseling for particular diseases or disabilities are reviewed in D. P. Agle, "Psychiatric Studies of Patients with Hemophilia and Related States," *Archives of Internal Medicine,* Vol. 114 (1962) 76–82; K. H. Halloran et al., "Genetic Counseling for Congenital Heart Disease," *The Journal of Pediatrics,* Vol. 88 (1976), 1054–56; and M. Giannini and L. Goodman, "Counseling Families During the Crisis Reaction to Mongolism," *American Journal of Mental Deficiency,* Vol. 67 (1963), 740–47. The concept of a "fear of knowledge" is developed from Abraham H. Maslow, *The Psychology of Science* (Henry Regnery, Chicago, 1969), pp. 17*ff.*

The possible biases in Sir Cyril Burt's work on twin studies and the inheritance of intelligence,* long considered classics in the field, were first challenged in Leon Kamin, *The Science and Politics of I.Q.* (John Wiley, New York, 1974). Full acceptance of data indicating self-deception (or, as Kamin believes, outright fraud) has come only grudgingly. (See especially, D. D. Dorfman, "The Cyril Burt Question: New Findings," *Science,* Vol. 201 (1978), 1177–78.)

Early sources for the concept that human frailty and mental illness were inherited include Abraham Myerson, *The Inheritance of Mental Diseases* (Williams and Wilkins, Baltimore, 1925) and Michael F. Guyer, *Being Well-Born* (Bobbs-Merrill, Indianapolis, 1916).

General sources for an introduction to the XYY question (technical references follow in Chapter 4) can be found in Tabitha Powledge, "The XYY Man: Do Criminals Really Have Abnormal Genes? *Science Digest,* Vol. 79 (1976), 33–38; and in Richard Roblin, "The Boston XYY Case," *Hastings Center Report,* Vol. 5 (1975), 5–8.

*Most notably, "The Genetic Determination of Differences in Intelligence: A Study of Monozygotic Twins Reared Together and Apart," *British Journal of Psychology,* Vol. 57 (1966), 151; and "The Evidence for the Concept of Intelligence," *British Journal of Educational Psychology,* Vol. 25 (1955), 167.

CHAPTER THREE

(pages 53–78)

In summarizing the story of genetic variation I relied on a number of primary sources. The classic ones include R. C. Lewontin and J. L. Hubby, "Amount of Variation and Degree of Heterozygosity in Natural Populations of Drosophilia," *Genetics,* Vol. 54 (1966), 595–609; H. T. Band and P. T. Ives, "Genetic Structure of Populations," *Evolution,* Vol. 17 (1963), 198–315; and H. Harris and D. A. Hopkinson, "Average Heterozygosity per Locus in Man: An Estimate Based on the Incidence of Enzyme Polymorphisms," *Annals of Human Genetics* (London), Vol. 36 (1972), 9–19.

The significance of this variation is discussed in two accessible sources: E. O. Wilson, "Evolutionary Biology Seeks the Meaning of Life Itself," *New York Times,* Nov. 27, 1977 (Op Ed page); and Gina Bari Kolata's award-winning series, "Population Genetics: Reevaluation of Genetic Variation," *Science,* Vol. 184 (1974), 452–54. The list of protagonists in this debate is, of course, partial. Notable "neutral" theorists include Thomas Jukes, University of California at Berkeley, and James Crow, University of Wisconsin; advocates of the adaptive value of variation ("selectionists") include Richard C. Lewontin, of Harvard University, and Francisco Ayala, of the University of California at Davis.

The interested reader may wish to review one or more technical studies that support these viewpoints, specifically T. Yamazaki and T. Maruyama, "Evidence of the Neutral Hypothesis of Protein Polymorphism," *Science,* Vol. 178 (1972), 55–57; and J. M. Thoday, "Non-Darwinian Evolution and Biological Progress," *Nature,* Vol. 255 (1975), 675–77. A thorough review of the selectionists' side can be found in R. C. Lewontin, *The Genetic Basis of Evolutionary Change* (Columbia University Press, New York, 1974). A dramatic supporting example of the adaptive value of a gene that is harmful in the homozygous state is found in D. C. Coleman, "Obesity Genes: Beneficial Effects in Heterozygous Mice," *Science,* Vol. 203 (1979), 663–65.

In representing the issue of consanguinity I tried to balance the widely held view that inbreeding increases the occurrence of otherwise rare recessive genetic disorders with the evidence that the *over-all* effect of such inbreeding on such indicators as mortality and incidence of cancer is much less certain. A case in point is afforded by the Old

Order Amish, in which the incidence of specific genetic diseases is increased (see V. A. McKusick, "Dwarfism in the Amish," *Bulletin of the Johns Hopkins Hospital,* Vol. 115 [1964] 306), while the life span is prolonged. (See R. F. Hamman et al., "Patterns of Mortality in the Old Order Amish: Preliminary Report," unpublished paper presented at the American Public Health Association Meeting, Los Angeles, October 16, 1978.) A much more comprehensive discussion can be found in W. J. Schull and J. V. Neel, *The Effects of Inbreeding on Japanese Children* (Harper & Row, New York, 1965), whose research is reported in the *American Journal of Human Genetics* over a twenty-year period. (See Vol. 10 [1958], 398–445, and Vol. 24 [1972], 425 *et seq.)*

The reader interested in a general treatment of this topic is referred to V. A. McKusick, *Human Genetics,* 2nd ed. (Prentice-Hall, Englewood Cliffs, N.J., 1969), pp. 164*ff.* A similarly authoritative review of the subject of sickle-cell anemia can be found in Harry Harris, *The Principles of Human Biochemical Genetics,* pp. 18 *et seq.*

The relationship of the so-called transplantation antigens to disease and immune responses constitutes an entirely discrete area of inquiry of its own. A bibliography available from Paul Terasaki at the University of California at Los Angeles School of Medicine deals with only *one* HLA antigen (B 27) and lists over 180 sources. Classic studies include P. I. Terasaki et al., "High Association of an HL-A Antigen, W27 *[sic],* with Ankylosing Spondylitis," *New England Journal of Medicine,* Vol. 288 (1973), 704–7. An excellent general reference is "HLA Disease Susceptibility: A Primer," *New England Journal of Medicine,* Vol. 297 (1977), 1060–63.

Specific references for the material used in the text on HLA include P. Rubinstein, N. Suciu-Foca and J. F. Nicholson, "Genetics of Juvenile Diabetes Mellitus," *New England Journal of Medicine,* Vol. 297 (1977), 1036–40; 1976; I. M. Roitt et al., "HLA DRw4 and Prognosis in Rheumatoid Arthritis," *Lancet,* Vol. 1 (1978), 990–92; and S. Vslin and J. F. Fries, "Striking Prevalance of Ankylosing Spondylitis in 'Healthy' W27 *[sic]* Positive Males and Females," *New England Journal of Medicine,* Vol. 293 (1975), 835–38. For an extremely readable, nontechnical discussion of the relevance of findings like these to occupational liability to disease, .see Tabitha Powledge, "Screening and Occupational Disease," *New Scientist,* Sept. 2, 1976, pp. 487–88. An example of the rapidity (and laxity) with which these associations become translated by the press into causal links can be found in "New Evidence that Arthritis Is 'Inherited,'" *San Francisco Chronicle,* June 2, 1978, p. 22.

For material on the possible association between gene(s) on the X chromosome and various forms of manic depression I relied on F. J. Cadoret and G. Winokur, "Genetics of Affective Disorders," in *Psychiatry and Genetics,* eds. M. A. Sperber and L. F. Jarvik (Basic Books, New York, 1976) pp. 66*ff.* (The specific link between responsiveness to treatment and family history referred to in the text is based on J. Mendlewicz, R. R. Fieve and F. Stallone, "Relationship Between the Effectiveness of Lithium Therapy and Family History," *American Journal of Psychiatry,* Vol. 130 [1973], 1011–17.) A popular treatment of the genetics of manic depression can be found in Aubrey Milunsky, *Know Your Genes* (Houghton Mifflin, Boston, 1977), pp. 232*ff.*

The points about the genetic hypotheses for schizophrenia and affective disorders are supported by R. J. Wyatt et al., "Low Platelet Monoamine Oxidase and Vulnerability to Schizophrenia," *Modern Problems in Pharmacopsychiatry,* Vol. 10 (1975), 38–56; and C. K. Cohn et al., "Reduced Catechol-O-Methyl Transferase Activity in Red Blood Cells of Women with Primary Affective Disorder," *Science,* Vol. 170 (1970), 1323–24. A synthetic view can be found in D. F. Horrobin, "Schizophrenia: Reconciliation of the Dopamine, Prostaglandin and Opioid Concepts and the Role of the Pineal," *Lancet,* Vol. I (1979), 529–31. Three genetic forms of depression are described in M.A. Schlesser, G. Winokur and B.M. Sherman, "Genetic Subtypes of Unipolar Primary Depressive Illness Distinguished by Hypothalamic-Pituitary-Adrenal Axis Activity," *Lancet* Vol. 1 (1979), 739-41. Contemporary studies of this subject can be found by keying to authors like David Rosenthal, Irving I. Gottesman, Leon Eisenberg, Leonard L. Heston, and George Winokur, most or all of whom may be expected to continue their researches in the genetics of mental illness. In discussing the novel hypothesis of heightened susceptibility of the carriers of DNA repair gene defects, I relied on an elegant essay by Jean L. Marx entitled "DNA Repair: New Clues to Carcinogenesis," *Science,* Vol. 200 (1978), 518–21. General discussion of the genetics of Huntington disease is available in *Psychiatry and Genetics,* pp. 4*ff.,* while discussion of the pros and cons of preemptive testing for this disorder can be found in M. Hemphill, "Pretesting for Huntington's Disease," *Hastings Center Report,* Vol. 3 (1973), 12–13; and in W. Gaylin, "Genetic Screening: The Ethics of Knowing," *New England Journal of Medicine,* Vol. 286 (1972), 1361. (The latter editorial accompanies the Klawans article referred to in the text).

I relied on two sources in discussing the dilemmas of screening for Mendelian disorders wherein the health significance of the gene defect is uncertain: R. W. Erbe, "Screening and Genetic Counseling in Mendelian Disorders," *Birth Defects Original Article Series,* Vol. 10, No. 6 (1974), 85–100 (hyperlipidemias); and J. S. Popkin et al., "Is Hereditary Histidinemia Harmful?" *Lancet,* Vol. 1 (1974), 721–22 (see also editorial on p. 219). Heart-disease susceptibility and genetic makeup are linked in J. L. Goldstein et al., "Hyperlipidemia in Coronary Heart Disease," *Journal of Clinical Investigation,* Vol. 52 (1973), 1544–68.

The works of John Money illuminate the issue of psychosocial versus chromosomal determinants of sex and gender. See J. Money and A. A. Ehrhardt, *Man and Woman, Boy and Girl* (Johns Hopkins University Press, Baltimore, 1972); and J. Money, *Sex Errors of the Body: Dilemmas, Education and Counseling* (Johns Hopkins Press, Baltimore, 1968). For a particularly sensitive discussion of this issue, see R. J. Stoller, "Genetics, Constitution and Gender Disorder," *Psychiatry and Genetics,* pp. 41*ff.*

I gave weight to the studies of Lee Willerman and Sandra Scarr and their co-workers in critiquing the view that intelligence varies according to genetic differences among races or social classes. See L. Willerman, A. F. Naylor and N. C. Myrianthropoulos, "Intellectual Development of Children from Interracial Matings: Performance in Infancy and at Four Years," *Behavior Genetics,* Vol. 4 (1974), 83–88; and Sandra Scarr Salatapetak, "Race, Social Class and IQ," *Science,* Vol. 174 (1971), 1292–1305.

Measuring the "proportion of white admixture" in blacks, or estimating and/or applying "heritability" figures for presumptive between-group differences in genetic makeup has a wide literature. Among the sources used are J. M. Thoday, "Limitations to Genetic Comparisons of Populations," *Journal of Biosocial Science,* Suppl. 1 (1969), 3–14; J. Adams and R. H. Ward, "Admixture Studies and the Detection of Selection," *Science,* Vol 180 (1973), 1137–43 (admixture); R. C. Lewontin, "The Analysis of Variance and the Analysis of Causes," *American Journal of Human Genetics,* Vol. 26 (1974), 400–411; and N. E. Morton, "Analysis of Family Resemblance, I. Introduction," *American Journal of Human Genetics,* Vol. 26 (1974), 318–20 (heritability).

An extremely accessible treatment of the whole debate is M. W. Feldman and R. C. Lewontin, "The Heritability Hang-up," *Science,* Vol. 190 (1975), 1163–68.

References to Arthur Jensen in the text are drawn from two of his major works: "How Much Can We Boost IQ and Scholastic Achievement?" *Harvard Education Review*, Vol. 39 (1969), 1–123; *Educability and Group Differences* (Harper & Row, New York, 1973); and from his Statement to the Senate Select Committee on Education (Feb. 24, 1972). Equally weighty rejoinders and reviews of the implications of the hereditarian position can be found in Daniel Bell, "On Meritocracy and Equality," *The Public Interest*, No. 29 (Fall 1972), pp. 29–68; Christopher Jencks, *Inequality: A Reassessment of the Effect of Family and Schooling in America* (Basic Books, New York, 1972); and James Coleman, "Equal Schools or Equal Students?" *The Public Interest*, No.4 (Summer 1966).

The interested reader is referred to two extremely readable sources for further elucidation of the limitations of both environmentalist and hereditarian schools of thought: S. Scarr and R. A. Weinberg, "Attitudes, Interests and IQ," *Human Nature*, Vol. 1 (April 1978), 29–36; and J. Loehlin, G. Lindzey and J. N. Spuhler, *Race Differences in Intelligence* (W. H. Freeman, San Francisco, 1975).

CHAPTER FOUR

(pages 79–99)

The scope of screening efforts for genetic disease is surveyed in A. J. Bennett, "New England Regional Newborn Screening Program," *New England Journal of Medicine*, Vol. 297 (1977), 1178–84. Representative technical refinements are exemplified by J. Mee et al., "Rapid and Quantitative Blood Amino Acid Analysis by Chemical Ionization for Mass Spectrometry," *Biomedical Mass Spectrometry*, Vol. 4 (1977), 178–81; and reviewed in M. Lappé and R. O. Roblin, "Newborn Genetic Screening as a Concept in Health Care Delivery: A Critique," *Birth Defects Original Article Series*, 10(6) (1974), 5–16. A comprehensive picture of the field is available in H. L. Levy, "Genetic Screening," *Advances in Human Genetics*, Vol. 4 (1973), 1–104.

Two accessible sources that highlight the problems posed by the extension of genetic theory into society are Philip Reilly, *Genetics, Law, and Social Policy*, and Daniel Callahan, *Tyranny of Survival* (Simon and Schuster, New York, 1975). A general medical overview of the history of screening is found in J. M. G. Wilson and G. Jungner, *Principles and Practice of Screening for Disease* (World Health Organization, Geneva, 1968).

The report generally credited with triggering the intense interest in XYY males is P. A. Jacobs et al., "Aggressive Behavior, Mental

Subnormality and the XYY Male," *Nature,* Vol. 208 (1965), 1351–53. Popular treatments that followed included M. A. Telfer, "Are Some Criminals Born That Way?" *Think,* Vol. 34 (1968) 24–28; and Anonymous, "Congenital Criminals," *Newsweek,* Vol. 75 (1970), 98–99.

Different research groups appeared to generate conflicting results, with a Danish group headed by Johannes Nielsen documenting behavioral deviance as a common pattern of developing XYY males (see J. Nielsen et al., "Childhood of Males with the XYY Syndrome," *Journal of Autism and Childhood Schizophrenia,* Vol. 3 [1973], 5–26); and a Colorado-based one led by Martha F. Leonard showing that XYY males were often indistinguishable behaviorally from their normal peers (see M. F. Leonard et al., "Early Development of Children with Abnormalities of the Sex Chromosomes: A Prospective Study" *Pediatrics,* Vol. 54 [1974], 208–12.)

Some reports failed to find even minor behavioral or intelligence deficits (see *Clinical Genetics,* Vol. 5 (1974), 387–94; and *Pediatrics,* Vol. 48 [1971], 583–94). The most systematic study concluded that the adverse developmental impact of either an extra X or Y chromosome was comparable—that mental retardation was the common denominator in incarcerated XXY or XYY males—and therefore discouraged additional screening intended to prospectively identify at-risk individuals (H. A. Witkin et al., "Criminality in XYY and XXY Men," *Science,* Vol. 193 [1976], 547–55).

Newborn screening for chromosomal abnormalities was most intensive in the period between 1968 and 1976. Representative centers of activity included Boston (Park S. Gerald); Winnipeg (John L. Hamerton); New Haven (Frank H. Ruddle); Edinburgh (Patricia Jacobs); London, Ontario (Frederick Sergovich); and Arhus, Denmark (Johannes Nielsen).

The Law Enforcement Assistance Administration report cited in the text can be found in the *LEAA Annual Report* for 1971, in which $79,900 is identified as going to the Behavioral Science Foundation of Cambridge, Mass., to ascertain "the feasibility of using fingerprints as a rough index to identify individuals who are most likely to exhibit chromosomal aberrations."

For reviews of the social and behavioral hypotheses for deviance and the constitutional and legal implications of XYY identification, see E. B. Hook, "Behavioral Implications of the Human XYY Genotype," *Science,* Vol. 179 (1973), 139–44; and P. N. Brown, "Guilt by Physiology: The Constitutionality of Tests to Determine Predisposition to Violence," *Southern California Law Review,* Vol. 48 (1974), 489–570, respectively. (Popular treatments of the XYY problem are given in the notes for Chapter 2).

The psychological impact of screening on the subject has focused on Tay-Sachs and sickle-cell anemia screening. The impact of Tay-Sachs screening is reported in M. D. Kuhr, "Doubtful Benefits of Tay-Sachs Screening," *New England Journal of Medicine,* Vol. 292 (1975) 371; B. Childs et al., "Tay-Sachs Screening: Social and Psychological Impact," *American Journal of Human Genetics,* Vol. 28 (1976), 550–58; and in a sensitive paper by C. L. Clow and C. R. Scriver, "Knowledge About and Attitudes Toward Genetic Screening Among High-School Students: The Tay-Sachs Experience," *Pediatrics,* Vol. 54 (1977), 86–91. A definitive text is M. M. Kaback, D. L. Rimoin and J. S. O'Brien, *Tay-Sachs Disease: Screening and Prevention* (Alan Liss, New York, 1977).

The stress of sickle-cell screening is documented by R. M. Antley, "Responses to Genetic Counseling of Positive Families after Sickle-Cell Screening," *American Journal of Human Genetics,* Vol. 25 (1973), 12A; S. Kumar et al., "Anxiety, Self-Concept, and Personal and Social Adjustments in Children with Sickle-Cell Anemia," *The Journal of Pediatrics,* Vol. 88 (1976), 859–63; and M. L. Hampton, "Sickle-Cell Disease and Sickle-Cell Trait Screening Called Potentially Harmful," *Medical Tribune,* June 20, 1973. Stigmatization is reviewed in R. H. Kenen and R. M. Schmidt, "Stigmatization of Carrier Status: Social Implication of Heterozygote Genetic Screening Programs," *American Journal of Public Health,* Vol. 68 (1978), 1116–20. The belief that sickle-cell trait (not anemia) was harmful is documnted in D. B. Kellon et al., "Physicians' Attitudes About Sickle-Cell Trait," *Journal of the American Medical Association,* Vol. 227 (1974), 71–72; the study cited in the text was reported in *Genetic Screening: Programs, Principles and Research* (National Academy of Science, Washington, D.C., 1975).

Critical efforts to resolve the issue include: "The S-Hemoglobinopathies: An Evaluation of Their Status in the Armed Forces," *NAS–NRC Document,* Feb. 19, 1973, Government Printing Office; and M. S. Kramer, Y. Rooks and H. Pearson, "Growth and Development in Children with Sickle-Cell Trait: A Prospective Study of Matched Pairs," *New England Journal of Medicine,* Vol. 299 (1978), 686–89. Both downplayed the previously reported adverse effects of sickle-cell trait. The definitive report that documented the essential innocuousness of sickle-cell trait was P. Heller, W. R. Best, R. B. Nelson and J. Bechtel, "Clinical Implications of Sickle-Cell Trait and Glucose-6-phosphate Dehydrogenase Deficiency in Hospitalized Black Male Patients," *New England Journal of Medicine,* Vol. 300 (1979), 1001–5.

Evidence of discrimination is cited in Reilly (pp. 72*ff.*) and exemplified by the Job Corps program which advocated low-stress jobs for sickle-cell–trait carriers (J. Fielding et al., "A Coordinated Sickle-Cell Program for Economically Disadvantaged Adolescents," *American Journal of Public Health,* Vol. 88 [1974], 427–31.)

The expansion of screening into occupational illness or newborns potentially at risk for lung disease is discussed in Calabrese, *Pollutants and High Risk Groups,* pp. 55*ff.* The medical consensus appears to be that screening is not now justified for either the *Pi* phenotypes or aryl-hydrocarbon hydroxylase, a presumed causal factor in lung-cancer susceptibility (J. O. Morse et al., "Relation of Protease Inhibitor Phenotypes to Obstructive Lung Diseases in a Community," *New England Journal of Medicine,* Vol. 296 [1977], 1190–94; and B. Paigen et al., "Questionable Relation of Aryl Hydrocarbon Hydroxylase to Lung-Cancer Risk," *New England Journal of Medicine,* Vol. 297 [1977], 346–50 respectively).

Skeptics of PKU screening include S. P. Bessman and J. P. Swazey, "Phenylketonuria: A Study of Biomedical Legislation," in *Human Aspects of Biomedical Innovation,* eds. E. Mendelsohn, J. P. Swazey and I. Taviss (Harvard University Press, Cambridge, 1971), pp. 49–76. Advocates of the same period are exemplified by G. C. Cunningham's view in "Phenylketonuria Testing—Its Role in Pediatrics and Public Health," *Clinical Laboratory Science,* Vol. 2 (1971), 44–101. The limited predictive value of PKU testing cited in the text is drawn largely from R. S. Galen and S. R. Ganboli, *Beyond Normality: The Predictive Value and Effectiveness of Medical Diagnosis* (Wiley & Sons, New York, 1975), pp. 54–59. Perhaps the best balanced article on the subject in David Yi-Yung Hsia, "A Critical Evaluation of PKU Screening," *Hospital Practice,* Vol. 6 (1971), 101–12.

Detailed reviews of PKU screening are covered in W. E. Know, "Phenylketonuria," in *The Metabolic Basis of Inherited Disease,* eds. J. B. Wyngarden and D. S. Frederickson (McGraw-Hill, New York, 1972), pp. 265*ff.;* N. A. Holtzman, E. D. Mellits and C. Kallman, "Neonatal Screening for Phenylketonuria," *Pediatrics,* Vol. 53 (1974), 353–57; and C. R. Scriver, "Screening, Counseling and Treatment for Phenylketonuria: Lessons Learned—A Précis," in *Genetic Counseling,* eds. H. A. Lubs and F. de la Cruz (Raven Press, New York, 1977).

The assertion in the text that hypothyroid screening is more effective than PKU screening is supported by the first comprehensive review of the results of neonatal hypothyroid testing: D. A. Fisher et al., "Screening for Congenital Hypothyroidism: Results of Screening One

Million North American Infants," *Journal of Pediatrics*, Vol. 94 (1979), 700–5.

General reviews of genetic screening include the previously cited NAS study; D.N. Raine, "Inherited Metabolic Disease," *Lancet*, Vol. 2 (1974), 996–98, 996–97; and H. M. Nitowsky, "Prescriptive Screening for Inborn Errors of Metabolism: A Critique," *American Journal of Mental Deficiency*, Vol. 77 (1973), 538–50. The legislative issues are summarized in Reilly, *Genetics, Law and Social Policy*, pp. 37–121.

CHAPTER FIVE

(pages 100–112)

Philosophical implications of legislating the modifications of persons are discussed in N. N. Kittrie, *The Right to Be Different* (Johns Hopkins University Press, Baltimore, 1971). Ethical, social and legal issues are discussed in M. Lappé, R. O. Roblin and J. Gustafson, "Ethical and Social Issues in Screening for Genetic Disease," *New England Journal of Medicine*, Vol. 286 (1972), 1129–32.

Early warnings about overzealous legislation were sounded in 1972 and 1973 by C. F. Whitten, "Sickle-Cell Programming: An Imperiled Promise," *New England Journal of Medicine*, Vol. 288 (1973), 713–716; and B. J. Culliton, "Sickle-Cell Anemia: National Program Raises Problems as Well as Hopes," *Science*, Vol. 178 (1972), 283–85. By 1976 frank concern was experienced by B. J. Culliton, "Genetic Screening: The States May Be Writing the Wrong Kinds of Laws," *Science*, Vol. 191 (1976), 926–29.

The laws in question were reviewed extensively in 1975 by P. R. Reilly, "Genetic Screening Legislation," *Advances in Human Genetics*, Vol. 5 (1975), 319–76. Reform and "model" legislation were introduced in both California and Maryland (*California Statutes*, Chapter 1037 for 1977; *Maryland Statutes*, Article 43 Section 814 et seq. for 1976). A full discussion of the content of Maryland's law can be found in N. A. Holtzman, J. L. Lapides and G. J. Clarke, "Commission on Hereditary Disorders," *American Journal of Human Genetics*, Vol. 26 (1974), 523–24.

The federal legislation discussed in the text includes *Public Law 94–278:* Title IV, 90 Statutes, Section 401, "The National Sickle-Cell Anemia, Cooley's Anemia, Tay-Sachs and Genetic Diseases Act; *Public Law 92–94*, 86 Statutes, Section 136, "The National Sickle-Cell Anemia

Control Act of 1972; and *Public Law 92–414,* 86 Statutes, Section 605, The National Cooley's Anemia Control Act.

The study of counselors is that of James Sorenson, "Counselors: Self Portrait," *Genetic Counseling,* Vol. 1 (1973), 31–36; physician's attitudes are reported in *Genetic Screening: Programs, Principles and Research.*

An over-all view of normality and contrasting views of genetic deviation are presented in E. A. Murphy, "The Normal," *American Journal of Epidemiology,* Vol. 98 (1973), 403–11; and C. E. Dent, "Other Biochemical Abnormalities," in *Biochemical Screening in Relation to Mental Retardation,* ed. D. C. Cusworth (Pergamon Press, New York, 1971) pp. vii–viii. (C. E. Dent is the public-health researcher referred to in the text).

CHAPTER SIX

(pages 113–127)

The principal theme of this chapter is drawn from material originally appearing in M. Lappé, "Moral Obligations and the Fallacies of Genetic Control," *Theological Studies,* Vol. 33 (1972), 411–27. A contrasting view may be found in the writings of Joseph Fletcher, most notably, *The Ethics of Genetic Control* (Anchor Books, Garden City, N.Y., 1974).

Newspaper clips were taken from the San Jose *Mercury,* Dec. 11, 1977. Avery's letter was printed in "Letter from Oswald Avery to Roy Avery, May 17, 1943" in *Readings in Heredity,* ed. John A. Moore (Oxford University Press, New York, 1972) pp. 248–51. Mutual interactions of culture and language in developing scientific world views are discussed in *Language, Thought and Reality: Selected Writings of Benjamin Lee Whorf,* ed. J. B. Carroll (M.I.T. Press, Cambridge, Mass., 1956).

The reader interested in pursuing the concept of progress in evolution is referred to G. G. Simpson's famous essay, "The Concept of Progress in Organic Evolution," *Social Research,* Vol. 41 (1974), 28–51. A discussion of the stereotypic view of the Down syndrome child is presented in M. Lappé, "Genetic Knowledge and the Concept of Health," *Hastings Center Report,* Sept. 3, 1973, pp. 1–3. Malleability of the expression of this disability is documented in R. T. Bidder et al., "Benefit to Downs Syndrome Children Through Training Their Mothers," *Archives of Diseases in Childhood,* Vol. 50 (1975), 383–86.

An early human genetic text that overemphasizes both the likely power of genetic causation and the chromosome number in humans is Amram Scheinfeld, *Why You Are You: The Story of Heredity and Environment* (Association Press, New York, 1950). The actual discovery of the correct chromosome count is portrayed in M. J. Kottler, "From 48 to 46: Cytological Techniques, Preconceptions, and the Counting of Human Chromosomes," *Bulletin of the History of Medicine,* Vol. 48 (1965), 465–81.

Among those who challenged the orthodoxy were N. G. Martin, "No Evidence for a Genetic Basis of Tongue Rolling or Hand Clasping," *Journal of Heredity,* Vol. 66 (1975), 179–80; and A. P. Barnes and T. R. Mertens, "A Survey and Evaluation of Human Genetic Traits used in Classroom Lab Studies," *Journal of Heredity,* Vol. 67 (1976), 347–52. (At least one of their conclusions has since been challenged in the literature: E. L. Teng et al., "Handedness in a Chinese Population," *Science,* Vol. 193 [1976], 1148–50.) As an example of caution in advancing genetic hypotheses see the report by R. Mountain, X. Zwillich and J. Weil, "Hypoventilation in Obstructive Lung Disease," *New England Journal of Medicine,* Vol. 298 (1978), 521–25.

Reports of what appear to be exaggerated claims for the impact of genes on human lives can be found in Aubrey Milunsky, *Know Your Genes* (Houghton Mifflin, Boston, 1977); "Genetic Conditions" (California State Department of Education, 1977); *House of Representatives Report Number 498* (94th Congress, First Session for 1975; Government Printing Office); and J. A. F. Roberts, J. Chavez, and S. D. M. Court, "The Genetic Component in Child Mortality," *Archives of Diseases in Children,* Vol. 45 (1970), 33–38.

The more conservative figures presented in the text were drawn from N. Day and L. B. Holmes, "The Incidence of Genetic Disease in a University Hospital Population," *American Journal of Human Genetics,* Vol. 25 (1973), 237–46; and from Barton Childs, S. M. Miller and A. G. Bearn, "Gene Mutation as a Cause of Human Disease," in *Mutagenic Effects of Environmental Contaminants,* eds. H. E. Sutton and M. I. Harris (Academic Press, New York, 1972). Technical articles may be found in *Science,* Vol. 169 (1970), 495–97; and *Mutation Research,* Vol. 35 (1976), 387–91. An authoritative report which contains pertinent references to admissions due to genetic disease is Charles R. Scriver et al., "Genetics and Medicine: An Evolving Relationship," in *Health Care,* ed. Philip Abelson (American Association for the Advancement of Science, Washington, D.C., 1978) pp. 145–51.

The classic article describing the "four lethals" or their equivalents in terms of deleterious genes is a 1956 article by N. E. Morton, J. F. Crow and H. J. Muller, "An Estimate of the Mutational Damage in Man from Data on Consanguineous Marriages," *Proceedings of the National Academy of Sciences,* Vol. 42 (1956), 855–63. (Since 1956, some eighteen different studies have been published giving a median number of lethal recessive genes of 2.2—S. Sanharanarayanan, *Mutation Research,* Vol. 35 [1976], 387.)

The extrapolation of genetic theory made in *The Selfish Gene* is debated in the *Hastings Center Report,* December 1977, pp. 33–36, and August 1978, p. 4.

CHAPTER SEVEN

(pages 128–139)

I relied on three major sources in describing the key events in the history of ideas in genetics: L. S. Dunn, *A Short History of Genetics: Development of Some of the Main Lines of Thought* (McGraw Hill, New York, 1965); D. Hull, *Philosophy of Biological Science,* Foundations of Philosophy Series (Princeton University Press, Princeton, N.J., 1974); and J. A. Peters, ed., *Classic Papers in Genetics* (Prentice-Hall, Englewood Cliffs, N.J., 1959).

A comprehensive review of the varieties of thought in the nature-nurture debate can be found in the volume edited by L. Ehrman, G. S. Omenn, and E. Caspari, *Genetics, Environment and Behavior* (Academic Press, New York, 1972). A brilliant discussion of "buffering" in developmental systems can be found in C. H. Waddington, ed., "The Basic Ideas of Biology," in *Prolegomena: Towards a Theoretical Biology,* Vol. 1 (Aldine Publishers, Chicago, 1969), pp. 12*ff.* The role of heterozygosity in such buffering is discussed in I. M. Lerner, *Genetic Homeostasis* (Dover Publications, New York, 1970).

The classic paper on nature and nurture is J. B. S. Haldane, "The Interaction of Nature and Nurture," *Annals of Eugenics,* Vol. 13 (1946), 197–205. A contemporary view can be found in B. E. Ginsburg and W. S. Laughlin, "The Multiple Bases of Human Adaptability and Achievement: A Species Point of View," *Eugenics Quarterly,* Vol. 13 (1966), 240–47. Gunther Stent's views are set forth in a lecture he delivered at the Philosophy of Science Congress in London, Ontario, in 1974.

The hierarchal ranking of aggresivity in strains of mice was originally reported in J. P. Scott, "Genetic Differences in the Social Behavior of Inbred Strains of Mice," *Journal of Heredity,* Vol. 33 (1942), 11–15.

The companion piece was reported in B. E. Ginsburg and W. C. Allee, "Some Effects of Conditioning on Social Dominance and Subordination in Inbred Strains of Mice," *Physiological Zoology,* Vol. 15 (1942), 485–506.

The difficulties inherent in some anatomic studies where multiple genetic and environmental contributions are at work are exemplified in two studies: B. E. Eleftherious et al., "Cortex Weight: A Genetic Analysis in the Mouse," *Journal of Heredity,* Vol. 66 (1975), 207–12; and P. E. Polanyi, "Chromosomal and Other Genetic Influences on Birthweight Variation," in *Size at Birth,* Ciba Foundation Publication No. 27 (Elsevier, Netherlands, 1974), pp. 12*ff.*

Reviews on the relation of genetics to cancer span reports of dramatic single-gene associations (e.g., H. T. Lynch and A. R. Kaplan, "Cancer Concordance and the Hypothesis of Autosomal Dominant Transmission of Cancer Diathesis in a Remarkable Kindred," *Oncology,* Vol. 30 [1974], 210–16) to broad reviews of the complexity of gene-environment interactions (e.g., A. Knudson, "Heredity and Cancer in Man," *Progress in Medical Genetics,* Vol. 9 [1973], 113–32.) Two accessible overviews of the medical picture are E. Anderson, "Familial Susceptibility to Cancer," *CA: A Cancer Journal for Clinicians,* Vol. 26 (1976), 143–49; and Anonymous, "How Genes May Affect Susceptibility to Cancer," *New Scientist,* April 15, 1976, p. 129.

At issue is the translation of clear-cut animal studies (e.g., J. L. Bloom and D. S. Falconer, "A Gene with Major Effect on Susceptibility to Induced Lung Tumors in Mice," *Journal of the National Cancer Institute,* Vol. 33 [1964], 607–18) to humans. The discoverer of an analogous gene, C. R. Shaw, discusses the pro-screening perspective in "Carcinogen-Susceptibility Testing—Ethical Problem?" *Journal of the American Medical Association,* Vol. 232 (1975), 239–40.

Hypertension screening is at a similarly uncertain junction with relation to genetic causation. While there are single-gene–associated human conditions wherein hypertension is a concomitant (e.g., S. Karacada and T. Pirnar, "Hereditary Brachydactyly Associated with Hypertension," *Journal of Medical Genetics,* Vol. 10 [1973], 253–59), the consensus appears to favor multiple genetic and environmental overlays, with stress playing a mediating role (see R. Friedman and J. Iwai, "Genetic Predisposition and Stress-Induced Hypertension," *Science,* Vol. 193 [1976], 161–62.)

The experimental material cited in the text was drawn from J. Dupon et al., "Selection of Three Strains of Rats with Spontaneously Different Levels of Blood Pressure," *Biomedicine,* Vol. 19 (1973), 36–41. References to environmental factors include M. M. Kilcoyne, "Hypertension and Heart Disease in the Urban Community," *Bulletin of the New York Academy of Medicine*, Vol. 49 (1973), 501–10; and E. Harburg et al., "Sociological Stress, Suppressed Hostility, Skin Color, and Black-White Male Blood Pressure: Detroit," *Psychosomatic Medicine*, Vol. 35 (1973), 276–98. The role of genes is stressed in C. J. Maclean et al., "Genetic Studies on Hybrid *[sic]* Populations. III. Blood Pressure in an American Black Community," *American Journal of Human Genetics*, Vol. 26, (1974) 614-26; and E. Harburg et al., "Skin Color, Ethnicity and Blood Pressure. I. Detroit Blacks; and II. Detroit Whites," *American Journal of Public Health*, Vol. 68 (1978), 1177–1901.

For a discussion of the interrelationship between Western and Chinese thought and the generation of the inductive method in China, see Joseph Needham's classic *Science and Civilization in China*, Vol. 2 (Cambridge University Press, Cambridge, England, 1962), pp. 518*ff.*

CHAPTER EIGHT

(pages 140–155)

The classic source for the example of complex (i.e., polygenic) inheritance is Sewall Wright, "The Results of Crosses Between Inbred Strains of Guinea Pigs Differing in the Number of Digits," *Genetics*, Vol. 19 (1934), 537–51. A more recent review, which discusses the applicability of polygenic disorders to genetic screening is L. Ehrman and M. Lappé, "Screening for Polygenic Disorders," *Birth Defects Original Article Series*, Vol. 10 (1974), 107–22.

Readers interested in pursuing the variability inherent in the example of dominant inheritance used in the text are referred to N. L. Fienman and W. C. Yakowac, "Neurofibromatosis in Childhood," *Journal of Pediatrics*, Vol. 76 (1970), 339–41. At least one confounding factor in this and other genetic defects is the maternal environment (see M. Miller and J. G. Hall, "Possible Maternal Effect on Severity of Neurofibromatosis," *Lancet* Vol. 2 [1978], 1071–73.)

Two classic papers on the effect of stress operating through the maternal environment on later behavior in offspring are J. C. de Fries, "Prenatal Maternal Stress in Mice: Differential Effects on Behavior," *Journal of Heredity*, Vol. 55 (1964), 289–95; and W. R. Thompson, "The Effects of Prenatal Maternal Stress on Offspring Behavior in Rats," *Psychological Monographs*, Vol. 76 (1962), 38. Maternal stress as a birth-defect–producing agency in animals is discussed in L. P. Stean and L. A. Peer, "Stress as an Etiological Factor in the Development of

Cleft Palate," *Journal of Plastic and Reconstructive Surgery,* Vol. 18 (1956), 1–15 (cf. F. C. Fraser and D. Warburton, "No Association of Emotional Stress or Vitamin Supplement During Pregnancy to Cleft Lip or Palate in Man," *Journal of Plastic and Reconstructive Surgery,* Vol. 33 [1964], 395–97.) An over-all picture of the effects of maternal factors on development is discussed in D. W. Fulker, "Maternal Buffering of Rodent Genotype Response to Stress: A Complex Genotype–Environment Interaction," *Behavior Genetics,* Vol. 1 (1970), 119–24.

The story of complex causation in the production of cleft palate is drawn in part from H. Kalter, "Some Sources of Non-Genetic Variability in Steroid-Induced Cleft Palate in the Mouse," *Teratology,* Vol. 13 (1976), 1–10. A comprehensive source book for the whole field is J. G. Wilson and F. C. Fraser, eds., *Handbook of Teratology,* Vols. I and II (Plenum Press, New York, 1977).

The thalidomide episode is reviewed in "Britain's Great Thalidomide Cover-up," *Columbia Journalism Review,* May 1975. More general discussion of this and other birth defects can be found in L. V. Crowley, *An Introduction to Clinical Embryology* (Year Book Medical Publishers, Chicago, 1974) pp. 153*ff*. Embryological resilience in the face of insults such as the trypan-blue example in the text are discussed here and in *Science News,* May 3, 1975, p. 289; and in C. H. Waddington in *Prolegomena,* pp. 12–14.

Genetic heterogeneity for the disease models discussed in the text, specifically hemophilia, sickle-cell anemia, and cystic fibrosis (PKU is discussed in Chapter 2) was predicted by Henry Harris in 1953—*An Introduction to Biochemical Genetics*. The respective primary sources for this critical phenomenon, so crucial to the thesis of the book are D. F. Rogers, "The Genetic Basis of Variation in Factor 8 Levels Among Haemophiliacs," *Journal of Medical Genetics,* Vol. 8 (1971), 136–40; M. H. Steinberg et al., "Mild Sickle-Cell Disease," *Journal of the American Medical Association,* Vol. 224 (1973), 317–21; and H. Schwachtman, "The Heterogeneity of Cystic Fibrosis," *Birth Defects,* Vol. 8 (1972), 102–10, respectively. The individual described in the text was reported in R. R. Blanch and E. M. Mendoz, "Fertility in a Man with Cystic Fibrosis," *Journal of the American Medical Association,* Vol. 235 (1976), 1364–65.

The single-gene disorder that seemingly induces a stereotyped behavior was first described in M. Lesch and W. L. Nyhan, "A Familial Disorder of Uric Acid Metabolism and Central Nervous-System Function," *American Journal of Medicine,* Vol. 36 (1964), 561–66. The study that documents the apparent trainability of Lesch-Nyhan–affected boys is summarized in L. T. Anderson et al., "Lesch-Nyhan Disease," *Psychology Spectrum,* Vol. 10, (1975), 9.

Sources for studies on aggression and its genetic and environmental concomitants include K. M. J. Lagerspetz and K. Y. H. Lagerspetz, "The Expression of the Genes of Aggressiveness in Mice: The Effect of Androgen on Aggression and Sexual Behavior in Females," *Aggressive Behavior,* Vol. 1 (1975), 291–96; V. H. Denenberg, G. A. Hudgens and M. X. Zarrow, "Mice Reared with Rats: Modification of Behavior by Early Experience with Another Species," *Science,* Vol. 143 (1964), 380–81; and C. T. Randt, D. Z. Blizard and E. Friedman, "Early Life Undernutrition and Aggression in Two Mice Strains," *Developmental Psychobiology,* Vol. 8 (1975), 275–79.

The details of experimental corroboration (or lack thereof) for the link between Y chromosomes and aggressive behavior can be traced in M. K. Selmanoff et al., "Evidence for a Y Chromosomal Contribution to an Aggressive Phenotype in Inbred Mice," *Nature,* Vol. 253 (1975), 529–30; and M. K. Selmanoff, S. C. Maxson and B. E. Ginsburg, "Chromosomal Determinants of Intermale Aggressive Behavior in Inbred Mice," *Behavior Genetics,* Vol. 6 (1976), 53–69. For an exemplary treatment of the complex relationship between gene-environment interactions and criminal behavior, see Sarnoff A. Mednick and Karl O. Christiansen, eds., *Biosocial Bases of Criminal Behavior* (Gardner Press, New York, 1977) in which virtually all of the contributors are extremely cautious in deriving inferences about the relative influence of environment and heredity on criminality.

Alcohol and its genetic underpinnings are discussed at length in *Annals of the New York Academy of Sciences,* Vol. 197. The specific animal studies cited in the text are C. L. Randall and D. Lester, "Social Modification of Alcohol Consumption in Inbred Mice," *Science,* Vol. 189 (1975), 149–51 and Randall and Lester, "Alcohol Selection by DBA and C57BL Mice Arising from Ova Transfers," *Nature,* Vol. 255 (1975), 147–8. Other works suggesting a genetic component in human alcoholism include D. W. Goodwin et al., "Alcohol Problems in Adoptees Raised Apart from Alcoholic Biological Parents," *Archives of General Psychiatry,* Vol. 28 (1973), 238–43. A definite effect in men appears to be corroborated in more recent studies: D. D. Rutstein and R. L. Veech, "Genetics and Addiction to Alcohol," *New England Journal of Medicine,* Vol. 298 (1978), 1140–41.

For a feeling of the discordance between competing cultural and genetic models as major shapers of human destiny, compare L. L. Cavalli-Sforza and M. W. Feldman, "Cultural Versus Biological Inheritance: Phenotypic Transmission from Parents to Children," *American Journal of Human Genetics*, Vol. 25 (1973), 618–37; and E. O. Wilson's last chapter in *Sociobiology: The New Synthesis* (Harvard University Press, Cambridge, 1975); and *On Human Nature* (Harvard University Press, Cambridge, 1978), pp. 41*ff*.

CHAPTER NINE

(pages 156–175)

For a broadly ranging review of the place of free will versus determinism in scientific explanations of higher human functions, see the work of William James. (Cf. William Barrett, "Our Contemporary, William James," *Commentary*, December 1975, pp. 55–61; and writings by John W. Sutherland, "Beyond Behaviorism and Determinism," *Fields Within Fields*, No. 10 [Winter, 1973–74], pp. 31–47.)

I relied on Ernst Cassirer's *The Problem of Knowledge* (Yale University Press, New Haven, 1950) for much of my review of the history of ideas in biology. The insistence on the value of deterministic approaches in scientific methods is most forcefully argued in Ernest Nagel, *The Structure of Science: Problems in the Logic of Scientific Explanation* (Harcourt, Brace & World, New York, 1963); and François Jacob's *The Logic of Life*. Challenges to the orthodoxy can be found in A. Koestler and R. Smithies, *Beyond Reductionism* (Macmillan, New York, 1969); and a most readable article by Robert Walgate, "Breaking Through the Disenchantment," *New Scientist*, Sept. 18, 1975, pp. 667–69.

The interested reader is referred to the three-volume set entitled *Towards a Theoretical Biology*, ed. C. H. Waddington (Aldine Publishing, Chicago, 1969) for a constellation of views on the value of nonlinear, nonrigid approaches to discerning the functional relationships between genes and growth and form.

The linear approach to the "wiring diagram" modeling discussed in the text is exemplified by the work of R. W. Sperry (cf. "Mechanisms of Neural Maturation," in *Handbook of Experimental Psychology*, ed. S. S. Stevens) and Victor Hamburger (cf. "Emergence of Nervous Coordination," *Developmental Biology Supplement*, Vol. 2 (1968), 951–71). Persistent evidence for formal "blocks" of integrated neural circuitry exemplified in the text by the "silent stalking" behavior of the electrically-stimulated cat, are the exception to the rule of plasticity (cf. E. Preshansky and R. J. Bandler, "Midbrain-Hypothalamic Interrelationships in the Control of Aggressive Behavior," *Aggressive Behavior,* Vol. 1 [1975], 135–55.) The classic reports of single-gene effects on behavior include W. D. Kaplan and W. E. Trout III, "The Behavior of Four Neurological Mutants of Drosophila," *Genetics,* Vol. 61 (1969), 388–400; R. L. Collins and J. L. Fuller, "Audiogenic Seizure Prone *(Asp):* A Gene Affecting Behavior in Linkage Group VIII of the Mouse," *Science,* Vol. 162 (1968), 1137–39; and R. L. Sprott, "Passive-Avoidance Performance in Mice: Evidence of Single-Locus Inheritance," *Behavioral Biology,* Vol. 11 (1974), 231–37; and M. Bastock, "A Gene Mutation Which Changes a Behavior Pattern," *Evolution,* Vol. 10 (1956), 421–39.

The precision of invertebrate gene-related behavior is discussed in Sydney Brenner, "The Genetics of Behavior," *British Medical Bulletin,* Vol. 29 (1973), 269–71; a comparable general discussion of the difficulties in achieving precise genetic control in higher animals can be found in W. R. Thompson's discussion in *Genetics, Environment and Behavior,* pp. 209–14. A comprehensive bibliography is available for the experimental literature in R. L. Sprott and J. Staats, "Behavioral Studies Using Genetically Defined Mice—A Bibliography," *Behavior Genetics,* Vol. 5 (1975), 27–82; and an elegant synthetic view in T. Dobzhansky, "Of Flies and Men," in *Behavior Genetics: Method and Research,* eds. M. Manosevitz, G. Lindzey and D. D. Thiessen (Appleton Century Crofts, New York, 1969) pp. 58*ff.* An extremely readable popular version is Seymour Benzer's "From the Gene to Behavior," *Journal of the American Medical Association,* Vol. 218 (1971), 1015–26.

The whole issue of the plasticity of development in shaping neural pathways and brain structure was discussed at the Dahlem Conference in Berlin, 1977. Reports of the conference appear in *New Scientist,* March 3, 1977, and April 7, 1977; and in *Science,* Vol. 189 (1975), 207–9.

The original studies that provided evidence for the critical role of environmental stimulation in visual development were D. H. Hubel and T. N. Wiesel, "Receptive Fields of Cells in Striate Cortex of Very Young, Visually Inexperienced Kittens," *Journal of Neurophysiology,* Vol. 25 (1963), 994–1002; and T. N. Wiesel and D. H. Hubel, "Single-Cell Response in Striate Cortex of Kittens Deprived of Vision in One Eye," *Journal of Neurophysiology,* Vol. 25 (1963), 1003–17. Two excellent summaries of the state of the art are H. B. Barlow, "Visual Experience and Cortical Development," *Nature,* Vol. 258 (1975), 199–204; and Roger Lewin, "Cats' Brains are Controversial," *New Scientist,* Vol. 68 (1975), 457–58. The complexities intrinsic to development of binocular vision are revealed in the experiments by E. L. Smith III, et al., "Binocularity in Kittens Reared with Optically Induced Squint," *Science*, Vol. 204 (1979), 875–7.

The example of the "reeler" mouse used in the text is drawn from Shin-ho Ching, "The Brain of the 'Reeler' Mouse," *Nature*, Vol. 260 (1976), 14–15. For a related discussion see Anonymous, "Missing Protein Staggers Mice," *New Scientist*, March 11, 1976, p. 560; and "What Matters Make the Mind?" *New Scientist*, Jan. 1, 1976, p. 23. The dramatic role of biochemical gradients in nerve outgrowth is reviewed in a beautifully illustrated article: R. Levi-Montalconi and P. Calissano, "The Nerve Growth Factor," *Scientific American*, Vol. 240 (1979), 68–77.

Technical articles that provide the basis for my discussion of visual genetic control in albinism; in orientation visual acuity; and in ocular dominance are R. W. Guillery and J. H. Haas, "Genetic Abnormality of the Visual Pathways in a 'White' Tiger," *Science*, Vol. 180 (1973), 1287–88; B. N. Timney and D. W. Muir, "Orientation Anisotropy: Incidence and Magnitude in Caucasian and Chinese Subjects," *Science*, Vol. 193 (1967), 699–701; and P. Rakic, "Prenatal Genesis of Connections Subserving Ocular Dominance in the Rhesus Monkey," *Nature*, Vol. 261 (1971), 467–71, respectively.

Avian navigation, a prime example of gene-environment interactions, is discussed in S. T. Emlen, "The Stellar-Orientation System of a Migratory Bird," *Scientific American*, Vol. 233 (1975), 102–11.

Analogous technical articles from which the material on auditory genetic control were adduced include G. Gottlieb, "Species Identification by Avian Neonates: Contributory Effect of Prenatal Auditory Stimulation," *Animal Behavior*, Vol. 14 (1966), 282–90; and A. H. Riesen and D. E. Zilbert, "Behavioral Consequences of Variation in Early Sensory Environments," in *The Developmental Neuropsychology of Sensory Deprivation*, ed. A. H. Riesen (Academic Press, New York, 1974), pp. 211*ff*. Jean Piaget discusses the role of hereditary structures in child development in *The Origins of Intelligence in Children* (W. W. Norton, New York, 1963).

The general hypothesis of "deep," presumably gene-directed, grammatical structures lying behind the developmental makeup of human speech is explored in the writings of Noam Chomsky (most recently, *Language and Responsibility*, trans. John Viertel, [Pantheon Books, New York, 1979]). The extremely young age in which special perceptive skills emerge in infants (cf. P. D. Eimas, "Auditory and Phonetic Coding of the Cues for Speech: Discrimination of the [r–1] Distinction by Young Infants," *Perception and Psychophysics,* Vol. 18 [1975], 341–47), suggest strong genetic determination of phonetic boundaries, but does not fully substantiate the existence of genetic mechanisms for the control of innate grammars predicted by Chomsky.

CHAPTER TEN

(pages 176–190)

Wilhelm Roux's work and classic views of the integrated wholeness of organisms, beginning with Hans Driesch's *Science and Philosophy of the Organism,* are cited in Ernst Cassirer, *The Problem of Knowledge: Philosophy, Science and History since Hegel* (Yale Press, New Haven, 1950), pp. 178*ff*. Early "linear" views of embryogenesis are typified by the works of Paul Weiss (cf. "Self-Differentiation of the Basic Patterns of Coordination," *Comparative Psychology Monographs,* Vol. 17 [1941], 1–96). See pp. 28*ff* in Jacques Monod's *Chance and Necessity* (Knopf, New York, 1971) for a critique of holism. A contemporary advocacy of a reductionist viewpoint is found in the last chapter of E. O. Wilson's *Sociobiology*.

The field concepts of developmental potential are found in the writings of C. H. Waddington—see especially "The Genetical Control of Wing Development in Drosophila," *Journal of Genetics,* Vol. 41 (1940), 75–139; and "Fields and Gradients," in *Major Problems in Developmental Biology,* ed. M. Locke (Academic Press, New York, 1966). Specific sources used in discussing regional fields in limb development included L. Wolpert, "Positional Information and Morphogenetic Signals: An Introduction," in *Cell Interactions in Differentiation,* eds. M. Karkinen-Jaaskelanainen, L. Saxen and L. Weiss (Academic Press, New York, 1977); and S. V. Bryant and L. E. Hen, "Supernumary Limbs in Amphibians: Experimental Production in *N. viridescens* and a New Interpretation of Their Production," *Developmental Biology,* Vol. 59 (1976), 212–24. The controversial work on regeneration and cancer is by F. Seilern-Aspang and K. Kratochwil, "Induction and Differentiation of an Epithelial Tumor in the Newt (Triturus cristatus)," *Journal of Embryology and Experimental Morphology,* Vol. 10 (1962), 337–56.

An excellent overview of the field of morphogenesis can be found in M. Robertson, "Peaks and Valleys in Morphogenesis," *Nature,* Vol. 260 (1976), 394–95. The reader interested in a general source is referred to N. J. Berrill, *Developmental Biology* (McGraw-Hill, New York, 1971), pp. 389*ff*.

The dramatic plasticity of a unique kind of tumor cell described in the text was reported by B. Mintz and K. Illimensee, "Normal Genetically Mosaic Mice Produced from Malignant Teratocarcinoma Cells," *Proceedings of the National Academy of Science,* Vol. 72 (1975), 3585–89. For a general review, see B. Hogan, "Introducing Teratomas," *Nature,* Vol. 260 (1976), 466. An accessible treatment can be found in K. Illimensee and L. C. Stevens, "Teratomas and Chimeras," *Scientific American,* Vol. 240 (1979), 120–33.

A technical update on the "wobble" hypothesis can be found in Walter M. Fitch, "Is There Selection Against Wobble in Codon-Anticodon Pairing?" *Science,* Vol. 194 (1976), 1173–74.

Broadly ranging discussions of the problem of holistic approaches to knowledge can be found in a general article by A. MacIntyre and S. Gorovitz, "Toward a Theory of Medical Fallibility," in *Science, Ethics and Medicine,* eds. H. T. Engelhardt, Jr., and D. Callahan (Hastings Institute, Hastings-on-Hudson, N.Y. 1976); M. W. Feldman and R. C. Lewontin, "The Heritability Hang-up," *Science,* Vol. 190 (1975), 1163–68; Michael Polanyi, *The Tacit Dimension* (Doubleday, Garden City, N.Y., 1966); and Jacques Barzun, *Darwin, Marx and Wagner: A Critique of a Heritage* (Doubleday, Garden City, N.Y., 1958).

The integrated processes by which "deep" structures and environmental stimulation interact to permit pattern recognition and neuronal development are explored in C. C. Goren, M. Sary and P. Y. K. Wu, "Visual Following and Pattern Discrimination of Face-like Stimuli by Newborn Infants," *Pediatrics,* Vol. 56 (1975), 544–49; and more generally in L. B. Cohen and P. Salapatek, eds., *Infant Perception* (Academic Press, New York, 1975) especially, Volume 2, "Perception of Space, Speech and Sound." An excellent general review that provides a balanced viewpoint of the relative role of genetic propensities, and ranges of sensitivity and experiential information is I. Grobstein and K. L. Chow, "Receptive Field Development and Individual Experience," *Science,* Vol. 190 (1975), 352–58.

Piaget's observations and the general tone of the whole movement to reexamine the reductionist philosophical core of contemporary science are found in A. Koestler and R. Smithies, eds., *Beyond Reductionism* (Macmillan, New York, 1969), especially, J. Piaget and R. Inhelder, "The Gaps in Empiricism," pp. 157*ff.* The place of modern physics in this movement is eloquently discussed in J. Bronowski, *The Ascent of Man* (Little, Brown, Boston, 1973), pp. 364*ff.*

For an example of the cultural dissonance in receptivity to the place of quantum-theory-like explanations in biology, the reader is referred to

E. R. John's observations in "Russians and Americans Gather to Talk Sociobiology," *Science,* Vol. 200 (1973), 631–33.

The reader interested in the interface between quantum-physical theory and philosophy is referred to Niels Bohr, *Atomic Physics and Human Knowledge* (Science Editions, New York, 1958); and Werner Heisenberg, *Physics and Philosophy* (Harper & Row, New York, 1968).

A definitive demonstration of the thesis of this Chapter, that chance can disturb the expression of genes in otherwise identical single cells, is reported in J. L. Spudick and D. E. Koshland, Jr., "Non-Genetic Individuality: Chance in the Single Cell," *Nature,* Vol. 262 (1976), 467–71.

CHAPTER ELEVEN

(pages 191–208)

The quotation is from Robert Walgate's essay, "Breaking Through the Disenchantment," *New Scientist,* Sept. 18, 1975, pp. 667*ff.*

The case against screening for alpha-l-antitrypsin–deficiency screening is developed in detail in Chapter 4. References used for this section include M. L. O'Brien, N. R. M. Buist and W. H. Murphey, "Neonatal Screening for Alpha-l-Antitrypsin Deficiency," *Journal of Pediatrics,* Vol. 92 (1978), 1006–10. General sources for further reading are H. L. Sharp, "Alpha-l-Antitrypsin Deficiency," *Hospital Practice,* May 1971, pp. 83–96; and F. Prieto, "Lung Disease and the Environment," *New Scientist,* March 3, 1976. The asbestosis-B27 link is hypothesized in J. A. Marchant et al., "The HL-A System in Asbestos Workers," *British Medical Journal,* Vol. 1 (1975), 189–91.

The position of those who see an expanded role for screening is exemplified by Barton Childs, "Prospects for Genetic Screening," *Journal of Pediatrics,* Vol. 87 (1975), 1124–32.

The relatively inconsequential role of genetics in obesity is discussed in a syndicated article by Jane E. Brody, science writer for *The New York Times:* "Science Finds New Fat Clues," *The Sacramento Bee,* March 8, 1979, p. D1. A technical study that refutes the existence of a significant correlation between presumably highly genetically controlled weight at birth and later obesity is reported in M. S. Dine et al., "Where Do All the Heaviest Children Come From?" *Pediatrics,* Vol. 63 (1979), 1–7. The source used in the text is T. J. Coates and C. E. Thoresen, "Treating Obesity in Children and Adolescents: A Review," *American Journal of Public Health,* Vol. 68 (1978), 143–49. A contrasting view advocating the susceptibility of polygenic disorders to simple gene analysis is found in H. Valtin, "Genetic Models in Biomedical Investigation," *New England Journal of Medicine,* Vol. 290 (1974), 670–75.

A heavy blow to the genetics-IQ controversy is struck by the article by French researchers showing that social class IQ differentials appear explainable on the basis of cultural, not genetic factors: N. Schiff et al., "Intellectual Status of Working-Class Children Adopted into Upper-middle Class Families," *Science,* Vol. 200 (1978), 1503–4.

The views of distributive justice discussed in the text are drawn in large part from the writings of John Rawls; see especially his *A Principle of Justice* (Belknap Press, Cambridge, Mass., 1972). The quotation from Wilson on the desirability of universal work standards was cited unfavorably in Tom Alexander, "OSHA's Ill-Conceived Crusade Against Cancer," *Fortune,* July 3, 1978, pp. 86–90. Sources on typological thinking include the previously cited writings of Ernst Mayr and the text by J. L. Fuller and W. R. Thompson, *Behavior Genetics* (John Wiley, New York, 1960).

The Eastern perspective is developed in Joseph Needham's essay, "History and Human Values: a Chinese Perspective for World Science and Technology," *The Centennial Review,* Vol. 20 (1976), 1–35; John Gurley, "Capitalist and Maoist Development," *Monthly Review,* February 1971, pp. 15*ff;* and Urie Bronfenbrenner, "Child Watching in a Chinese Classroom," *New York Times,* Jan. 15, 1975, p. 95 (Op-Ed page).

Position statements on the desirability of applying genetic knowledge to eugenic aims range from the frank advocacy views of I. I. Gottesman and L. Erlenmeyer Kimling, "Prologue: A Foundation for Informed Eugenics," *Social Biology Supplement,* Vol. 18 (1971), 1–18, to the moderate ones of C. R. Scriver and his colleagues: C. R. Scriver, C. Laberge, C. L. Clow and F. Clarke Fraser, "Genetics and Medicine: An Evolving Relationship," *Science,* Vol. 200 (1978), 946–52. The openly skeptical position presented in the text is more fully developed in the chapter by M. Lappé, "Can Eugenic Policy Be Just?" in *The Prevention of Genetic Disease and Mental Retardation,* ed. A. Milunsky, (W. B. Saunders, Philadelphia, 1975), pp. 456–75.

The relationship between genetic predisposition and normative values for birth weight, hemoglobin status, iron requirements, blood pressure and stature is still in doubt. Optimum as well as normal values for each of these parameters are increasingly being proposed on the basis of yet-to-be-elucidated genetic differences between racial groups.

Key references include, R. L. Naeye et al., "Relation of Poverty and Race to Birth Weight and Organ and Cell Structure in the Newborn," *Pediatric Research,* Vol. 5 (1971), 17–28; C. L. Johnson and S. Abraham, "Hemoglobin and Selected Iron-Related Findings of Persons 1–74 Years of Age: United States, 1971–1974," *Advance Data,* No. 46, Jan. 26, 1979 (National Center for Health Statistics) pp. 1–11; R. K. Chandra, "Impaired Immunocompetence Associated with Iron Deficiency," *Journal of Pediatrics,* Vol. 86 (1975), 899–922; A. A. Tyroler and S. A. James, "Blood Pressure and Skin Color" (editorial), *American Journal of Public Health,* Vol. 68 (1978), 1170–72; and M. Garn and D. C. Clark,* "Problems in the Nutritional Assessment of Black Individuals," *American Journal of Public Health,* Vol. 66 (1976), 262–67.

The nutritional data alluded to in the text is from the *National Nutrition Survey in California, 1969.*

CONCLUSION

(pages 209–216)

The quotation is taken from John C. Loehlin, Gardner Lindzey and J. N. Spuhler, *Race Differences in Intelligence* (W. H. Freeman, San Francisco, 1975), probably the most approachable and complete text for the viewpoint that systematic genetic differences within groups contribute to differences in IQ scores. Among the most compelling counterstudies that isolate an environmental factor in intelligence and psychomotor performance is H. L. Needleman et al., "Psychological Performance of Children with Elevated Lead Levels," *New England Journal of Medicine,* Vol. 300 (1979), 689–94. The genetic model of schizophrenia used in the Soviet Union and its perils are discussed in Warren Reich, "The Spectrum Concept of Schizophrenia," *Archives of General Psychiatry,* Vol. 32 (1975), 489–97.

Bernard Davis' views are expressed in the issue of the *New York Academy of Sciences Annals* devoted to a review of ethical and scientific issues in genetics: M. Lappé and R. S. Morison, eds. "Ethical and Scientific Issues Posed by Human Uses of Molecular Genetics," *Annals of the New York Academy of Sciences,* Vol. 265 (1976), 109*ff.*

For further reading on the emerging role of social scrutiny into the purposes and directions of scientific inquiry and its perceived desirability, see the essays and letters collected in *Genetic Engineering:*

* Garn and Clark are the two researchers who advanced the genetic hypothesis for racial differences in nutrition alluded to in the text.

Evolution of a Technological Issue, Report prepared for the Subcommittee on Science, Research and Development of the Committee on Science and Astronautics, U.S. House of Representatives (Government Printing Office, Washington, D.C., 1974), and subsequent publications of this committee's findings on the recombinant DNA issue. For legal commentary, see I.S. Bass, "Governmental Control of Research in Positive Eugenics," *Journal of Law Reform,* Vol. 8 (Spring, 1974), 615–30.

For an excellent overview of this debate, see T. M. Powledge, "Dangerous Research and Public Obligation," *New York Times,* Feb. 15, 1977, p. 8. A protagonist of the view that would have scientific research scrutinized for its public effects is Salvadore E. Luria, "Biological Roots of Ethical Principles," in *Genetics and the Law,* eds. A. Milunsky and G. Annas (Plenum Press, New York, 1976) pp. 407–10; an antagonist is Arthur G. Steinberg, "The Social Control of Science, *Genetics and the Law,* pp. 301–10.

Evidence for a socioeconomic gradient in birth defects (the poor have more) is found in *Infant Mortality Rates: Socioeconomic Factors,* DHEW Publication No. (HSM) 72–1045, Tables 14 and 15, pp. 34–5.

The view that scientific inquiry should be subject to scrutiny for its possible societal and moral consequences *before* it is undertaken is developed more fully in M. Lappé, "Reflections on the Nonneutrality of Hypothesis Formulation," *Clinical Pediatrics,* Vol. 24 (1976), 56–63.

Index